HANDBOOK OF MECHANICAL ENGINEERING TERMS

HANDBOOK OF MECHANICAL ENGINEERING TERMS

K.K. Ramalingam
Professor and Head (Retd.)
Dept. of Automobile Engineering
Madras Institute of Technology
Anna University
Chennai-600025

PUBLISHING FOR ONE WORLD

NEW AGE INTERNATIONAL (P) LIMITED, PUBLISHERS
New Delhi • Bangalore • Chennai • Cochin • Guwahati • Hyderabad
Jalandhar • Kolkata • Lucknow • Mumbai • Ranchi
Visit us at **www.newagepublishers.com**

Published by New Age International (P) Ltd., Publishers
First Edition: 1996
Second Edition: 2009

Branches:

- 36, Malikarjuna Temple Street, Opp. ICWA, Basavanagudi, **Bangalore**. ✆ (080) 26677815
- 26, Damodaran Street, T. Nagar, **Chennai**. ✆ (044) 24353401
- Hemsen Complex, Mohd. Shah Road, Paltan Bazar, Near Starline Hotel, **Guwahati** ✆ (0361) 2543669
- No. 105, 1st Floor, Madhiray Kaveri Tower, 3-2-19, Azam Jahi Road, Nimboliadda **Hyderabad**. ✆ (040) 24652456
- RDB Chambers (Formerly Lotus Cinema) 106A, Ist Floor, S.N. Banerjee Road, **Kolkata** ✆ (033) 22275247
- 18, Madan Mohan Malviya Marg, **Lucknow**. ✆ (0522) 2209578
- 142C, Victor House, Ground Floor, N.M. Joshi Marg, Lower Parel, **Mumbai**. ✆ (022) 24927869
- 22, Golden House, Daryaganj, **New Delhi**. ✆ (011) 23262368, 23262370

ISBN: 978-81-224-2608-3

Rs. 695.00

C- 08-11-3232

Printed in India at Glorious Printer, Delhi
Typeset in-house

PUBLISHING FOR ONE WORLD

NEW AGE INTERNATIONAL (P) LIMITED, PUBLISHERS
4835/24, Ansari Road, Daryaganj, New Delhi - 110002
Visit us at **www.newagepublishers.com**

PREFACE TO THE SECOND EDITION

This revised edition of the book "HANDBOOK OF MECHANICAL ENGINEERING TERMS" is expected to be a source of information for students, teaching faculties and practicing engineers in the field of mechanical engineering.

About three hundred and fifty terms have been added in this revised edition. Every effort has been made to include the terms in current use.

Suggestions and comments are welcomed from the readers, which may help to improve this handbook.

K. K. Ramalingam

PREFACE TO THE FIRST EDITION

This handbook has been prepared to meet the need of an up-to-date authoritative, yet concise compilation of the salient terms associated with the broad field of Mechanical Engineering. This handbook is intended to serve a twofold purpose.

First, it concentrates on and presents a brief description of the most important currently used terms. It will give the reader a true and full understanding and background of the meaning of the words that sooner or later will be needed in any occupation or trade.

One point has been kept in mind, during the preparation of this handbook, that will, cause it to differ from other works of its class. This is accomplished by dividing the book into twenty six parts. These parts cover most of the spectrum of Mechanical Engineering. In each part, the various terms have been listed alphabetically.

Second, the book is available as in the general ready reference for the students and engineers of various courses and fields pertaining to Mechanical Engineering. It provides, in one convenient source, the brief explanations scattered in various text books, reference books handbooks and journals.

Every effort has been made to include terms in current use. All definitions have been expressed as clearly and simply as possible. Care has been exercised to avoid errors.

Suggestions and comments are heartily welcomed from the readers which may help to improve further editions of this handbook.

K. K. Ramalingam

CONTENTS

CHAPTER 1

GENERAL

ACCURATE – Without error within tolerances allowed, precise, correct, confirming exactly to standard.

ACHME THREAD – A screw thread having an included angle of 29° and largely used for feed screws on machine tools.

ACUTE ANGLE – An angle which is less than a right angle, 90°.

ADDENDUM – The portion of the tooth of a gear that extends from the pitch line to the outside.

ALIGN – To bring two or more components of a unit into correct positions with respect to one another.

ALLOWANCE – The intentional or desired difference between the maximum limits of mating parts to provide a certain class of fit.

ANGLE – The amount of opening or divergence between two straight lines that meet at a vertex or that intersect each other.

ANGLE OF THREAD – The included angle between the sides forming the groove of the screw thread.

ANNULUS – A figure bounded by concentric circles or cylinders (*e.g.*, a washer, ring, sleeve etc.).

ARC – A circular section of the circumference of a circle bounded by two equal radii.

ASSEMBLY – A unit that contains the parts that make up a mechanism or a machine.

AXIS – The line real or imaginary, which passes through the center of a body and about which the body would rotate if set revolving.

BACKLASH – The clearance or amount of movement between the tooth profiles of a pair or train of gears in mesh. Also refers to the looseness or lost motion between screw threads which have been badly worn.

BAFFLE – A device which slows down or diverts the flow of gases, liquid, sound etc.

BASIC SIZE – The theoretical or nominal standard size from which all variations are made.

BASTARD – Not standard, irregular. A bastard cut file is a rough cut file having coarse teeth than a second cut file.

BELL MOUTHED HOLE – A hole which is rounded or tapered slightly larger at one end or both ends and is not exactly cylindrical throughout its entire length.

BEVEL – Any surface not at right angle to the rest of the workpiece. If a bevel is at a 45° angle, it is frequently called a MITER.

BIMETALLIC STRIP – A strip of metal consisting of one metal (or alloy) in the top portion bonded to a different metal in the bottom portion. A straight strip becomes curved when heated.

BLIND HOLE – A hole which is made to a certain depth of a workpiece but does not pass through it.

BISECTING AN ANGLE – Dividing an angle into two equal parts.

BOND – The holding together of different parts.

BORE – The inside diameter of a cylinder, or a hole for a shaft. Also the operation of machining a circular hole in a metal workpiece.

BRUSH – Pieces of carbon or copper that make a sliding contact against the commutator or slip rings.

CAM – A plate or cylinder which transmits variable motion to a part of a machine by means of a follower.

CAP SCREW – A finished screw 5mm or larger, used for fastening two pieces together by passing the screw through a clearance hole in one part and screwing in into a tapped hole in the other.

CENTER – A fixed point about which the radius of a circle or an arc moves.

CENTER LINE – A line used on drawings to show the centers of objects and holes. The center line consists of alternate long and short dashes.

CHAMFER – To bevel or remove the sharp edge of a machined part.

CHECK VALVE – A valve which permits flow in one direction only.

CIRCULAR PITCH – The distance from the center of one gear tooth to the center of the next gear tooth measured on the pitch line.

CIRCUMFERENCE – A curved line forming a circle and the length of this line.

COIL SPRING – A spring steel wire wound in a spiral pattern.

COMMUTATOR – A number of copper bars connected to the armature windings but insulated from each other and from the armature.

CONVOLUTION – One full turn of screw.

CORE – The central or innermost part of an object.

COUNTER BORING – The operation of enlarging a portion of a hole for part of its depth and to a given diameter, as for the head of a fillister head screw.

COUNTER SINK – To cut or shape a depression in an object so that the head of a screw may set flush or below the surface.

CREST CLEARANCE – Defined on a screw form as the space between the top of a thread and the root of its mating thread.

CREST OF SCREW THREAD – The top surface joining the two sides of flanks of a thread.

CROWNED – A slight curve in a surface *e.g.*, on a roller or race way.

DEDENDUM – The depth of a gear tooth space below the pitch line or circle. Also, the radial distance between the pitch circle and the root circle, which also includes the clearance.

DIAPHRAGM – A flexible dividing partition separating two compartments.

DOUBLE FLARE – A flared end of the tubing having two wall thickness.

DOWEL – A pin, usually of circular shape like a cylinder, used to pin or fasten something in position temporarily or permanently.

DRIFT PIN – A round tapered steel pin used to align rivet holes so that the rivet will pass through the holes easily.

ECCENTRIC – A circle or cylinder having a different center from another coinciding circle or cylinder. Also, a device for converting rotary motion to reciprocating motion.

END PLAY – As applied to a shaft, the amount that the shaft can move backward and forward.

EYE BOLT – A bolt threaded at one end and bent to a loop at the other end.

FEATHER – A sliding key, sometimes called splint. Used to prevent a pulley, gear or other part from turning on a shaft but allows it to move lengthwise. The feather is usually fastened to the sliding piece.

FILLET – A concave surface connecting the two surfaces meeting at an angle.

FLANGE – A metal part which is spread out like a rim, the action of working a piece or part to spread out.

FLANK (Side of thread) – The straight part of the thread which connects the crest with the root.

FLARE – To open or spread outwardly.

FULCRUM – The pivot point of a lever.

FLUSH – When the surfaces of different parts are on the same level, they are said to be flush.

FLUTE – A straight or helical groove of angular or radial form machined in a cutting tool to provide cutting edges and to permit chips to escape and the cutting fluid to reach the cutting edges.

GATE VALVE – A common type of manually operated valve in which a sliding gate is used to obstruct the flow of fluid.

GEAR – A general term applied to types of toothed wheels, valve motion, pump works, lifting tackle and ropes.

GEARING – A train of gears or an arrangement of gears for transmitting motion in a machinery.

GIB – An angular or wedge like strip of metal placed between two machine parts, usually sliding bearings, to ensure a proper fit and provide adjustment for wear.

GLAND – A device to prevent the leakage of gas or liquid past a joint.

HAND WHEEL – Any of the various wheels found on machine tools for moving or positioning parts of the machine by hand feed, as the tailstock handwheel on a lathe.

HALF MOON KEY – A fastening device in a shape somewhat similar to a semicircle.

HELICAL GEAR – A gear in which the teeth are cut at some angle other than a right angle across the gear face.

HELICOIL – A thread insert used to repair worn or damaged threads. It is installed in a retapped hole to bring the screw thread down to original size.

HELIX – The curve formed by a line drawn or wrapped around a cylinder which advances uniformly along the axis for each revolution, as the thread on a screw or the flute on a twist drill. A helix is often called a spiral in the shop.

HELIX ANGLE OF A THREAD – The angle made by the helix of the thread at the pitch diameter line with a line at right angle to the axis.

HEXAGONAL NUT – A nut having six sides and shaped like a hexagon.

INVOLUTE GEAR TOOTH – A curved tooth generated by unwinding a string from a cylinder to form the curve.

JOURNAL – The part of a shaft or axle that has been machined or finished to fit into a bearing.

KEYS – Metal pieces of various designs that fit into a slot in a shaft and project above the shaft to fit into a mating slot in the center hole of a gear or pulley to provide a positive drive between the shaft and the gear or pulley.

KEYSEAT – The slot or recessed groove either in the shaft or gear, which is made to receive the key. Also, it is called a KEYWAY.

KNURL – A uniform roughened or checked surface of either a diamond, a straight or other pattern.

LAND – The top surface of a tooth of cutting tools, such as taps, reamers and milling cutters.

LEAD ANGLE – The angle of the helix of a screw thread or worm thread. It is the measure of the inclination of a screw thread from a plane perpendicular to the axis of the screw.

LEAD HOLE – A small hole drilled in a workpiece to reduce the feed pressure, aid in obtaining greater accuracy, and guide a large drill. Sometimes called PILOT HOLE.

LEAD OF THREAD – On a single threaded screw, the distance the screw or nut advances in one complete revolution.

LEFT HAND SCREW – One that screws into the mating part or advances when turned to the left or counter clockwise.

LIMITS OF SIZE – The minimum and maximum sizes permissible for specific dimensions.

LINE – A tube, pipe or hose which is used as a conductor of fluid.

LINKAGE – A movable connection between two units.

LOBE – The projecting part such as rotor lobe or the cam lobe.

LOCK NUT – A type of nut that is prevented from loosening under vibration. The locking action is accomplished by squeezing, gripping or jamming against the bolt threads.

LOOSE PULLEY – A pulley which turns freely on a shaft so that a belt can be shifted from the driving pulley to the loose pulley in order to stop a machine driven by an overhead belt drive.

MAJOR DIAMETER – On a straight thread, the diameter of the imaginary cylinder that just touches the crest of an external thread or the root of an internal thread.

MALE PART – The external part of any workpiece which fits into a hole, slot or groove of the mating part.

MANUAL VALVE – A valve which is opened or closed or adjusted by hand.

MESH – Engaging one part with another, as the teeth of one gear mesh with the teeth of a mating gear.

MICRON – One millionth of a meter or 0.039370 inch.

MINOR DIAMETER – On a straight thread, the diameter of the imaginary cylinder which just touches the root of an external thread or the crest of an internal thread.

MORSE TAPER – A self holding, standard taper largely used on drilling tools, drilling machine spindles, and some lathes.

MULTIPLE THREADED SCREW – A screw with two or more threads cut around the periphery of the workpiece to provide an increased lead with a specified pitch.

NUT – A metal fastener of square, hexagon or other shape, having an internal thread which screws onto a bolt, stud or arbor.

OFF CENTER – Not on the true center line or axis, offset, eccentric or inaccurate.

PEEN – The end of the head of a hammer opposite the face, such as ball, straight or cross peen, and used for peening or riveting.

PILOT – A guide at the end of the counter bore which fits freely into the drilled hole and align the body of the counterbore while cutting takes place.

PILOT SHAFT – A shaft positioned in or through a hole of a component as a means of aligning the components.

PILOT VALVE – A valve used to control the operation of another valve.

PINION – The smaller of the pair of gears regardless of the size or type.

PIPE THREAD – A 60° thread having flattened crest and roots which are cut on a taper. Pipe thread is used on piping and tubing.

PITCH – In screw threads, the distance from a point on one thread to a corresponding point on the next thread measured parallel to the axis. In the case of spur gears, indicates the size of the gear teeth and is correctly called diametral pitch.

PITCH DIAMETER – For screw threads, the diameter of an imaginary cylinder, the surface of which would pass through the threads at such points that would make the width of the groove and width of the land equal to one half the pitch.

PLAY – The movement between two components.

PULLEY – A wheel having a plain or V groove rim over which a belt runs for the transmission of power from one shaft to another.

QUILL – A hollow shaft that revolves on a solid shaft carrying pulleys, gears or clutches. When the clutch is closed, the quill and shaft revolve together.

RACK – A straight metal strip having teeth that mesh with those of a gear to convert rotary into reciprocating motion or just the opposite.

RATCHET – A gear with triangular shaped teeth to be engaged by a pawl which gives it intermittent motion or locks it against backward movement.

RECESS – A groove cut below the normal surface of a workpiece.

RIGHT HAND THREAD – A screw thread which advances into the mating part when turned clockwise or to the right.

RIVET – A one piece fastener consisting of a head and a body and used for fastening two or more pieces together by passing the body through a hole in each piece and then forming a second head on the body end. It cannot be removed except by taking off the head.

SCREW – A helix formed or cut on a cylindrical surface which may advance along the axis to the right or left. The helix may be single or multiple.

SCREW THREAD – A ridge of uniform section or shape in the form of a helix on the external or internal surface of a cylinder, or in the form of a conical spiral on the external or internal surface of a cone.

SET SCREW – Usually a hardened steel screw having either no head or a square head and with various degrees of points or ends to lock or tighten adjustable machine parts in position on a shaft.

SHOULDER SCREW – A screw having two or more diameters or shoulders and commonly used for supporting levers and other machine parts that have to operate freely.

SINGLE THREAD – A screw thread cut around a cylinder having a single start in which the lead is equal to the pitch.

SOCKET HEAD – Screw head having a hexagonal or other form of recessed socket in the head so that the screw can be turned with a wrench or key, as a hexagon key.

SPLINE – Slot or groove cut in a shaft or bore, a splined shaft onto which a hub, wheel, gear etc. with matching splines in its bore is assembled so that the two must turn together.

SPOOL VALVE – A hydraulic directional control valve in which the direction of the fluid is controlled by the means of a grooved cylindrical shaft (spool).

SPRING – An elastic device which yields under stress or pressure but returns to its original state or position when the stress or pressure is removed.

SPUR GEAR – A toothed wheel having external radial teeth.

SQUARE THREAD – A form of screw thread in which the cross–section of the thread forms a square, making the width of the thread equal to the space between the threads.

STEP BLOCK – A block of steel or cast iron having a series of steps and used for supporting the ends of machine clamps when clamping work to the table.

STOPS – Devices attached to the movable parts of a machine tool to limit the amount of travel.

STUD – A rod having thread on both ends.

STUFFING BOX – A chamber having manual adjustment device for sealing.

TAPER – A shaft or hole that gets gradually smaller toward one end.

TAPER PINS – Steel pins used for locating and holding the machine parts in position on a shaft.

T-BOLT – A threaded bolt having a square or rectangular end which fits into the T slot of a machine table for clamping workpieces.

TEMPLATE – A flat pattern or guide plate usually made from sheet metal and used as a gauge or guide when laying out, drilling, forming in a machine or filing irregular shapes on metal pieces.

THUMB SCREW – A type of screw having a winged or knurled head for turning by hand when a quick and light clamping effect is desired.

TOLERANCE – A fractional allowance for variation from the specifications.

T-SLOT – A recessed or undercut slot made with a special T shaped cutter in the tables of machine tools to receive the square head of a T bolt for clamping workpiece.

U-BOLT – An externally threaded fastener bent in the shape of the letter U and with both ends threaded.

VALVE – Any device or arrangement used to open or close an opening to permit or restrict the flow of a liquid, gas or vapour.

V-BLOCKS – Square or rectangular shaped blocks of steel that are usually hardened and accurately ground. These have 90° V groove through the center and are provided with clamps for holding round workpiece for laying out, drilling, milling etc.

VISE – A mechanical device of many designs and sizes in which work pieces are clamped for hand or machine operations.

V-WAYS – The top of the bed of a lathe, planer or other machine tool which acts as bearing surface for aligning and guiding the moving parts such as the carriage of a lathe.

WORM – A threaded cylinder which meshes with and drives a worm gear, the thread being specially designed to mate with the teeth in the worm gear.

WORM GEARS – Gears with teeth cut at an angle to be driven by a worm. The teeth are usually cut out with a hob to fit the worm.

CHAPTER 2

MATERIALS

ABRASIVE – A natural or artificial material such as sand stone, emery, aluminium oxide or silicon carbide.

ACID – A chemical term to define a material which gives an acid reaction.

ADDITIVES – Chemicals added to oil or fuel to increase its effectiveness and obtain desirable qualities.

ADHESIVES – Materials or compositions that enable two surfaces to join together. An adhesive is not necessarily a glue, which is considered to be a sticky substance, since many adhesives are not sticky.

AGGREGATE – Small particles such as powders that are used for powder metallurgy, that are loosely combined to form a whole, also sand and rock as used in concrete.

ALLOTROPIC METALS – Metals which exist in one lattice form over a range of temperature, but at a certain temperature the lattice form changes to another type which is stable over another temperature range.

ALLOY – A substance having metallic properties and is composed of two or more chemical elements, of which at least one is a metal.

ALLOYING ELEMENTS – Elements either metallic or non-metallic added intentionally to the base metal, to make a marked change in the properties of the base metal and to secure certain desirable properties.

ALLOY STEEL – Steel containing significant quantities of alloying elements (other than carbon and the commonly accepted amounts of manganese, silicon, sulphur and phosphorus) added to effect changes in mechanical and physical properties.

ALNICOS – Alnicos materials are composed mainly of aluminium, nickel, cobalt and iron. Some include additions of copper and titanium. They are high-coercive force, high magnetic energy alloys.

ALOXITE – Artificial abrasive material used in the manufacture of grinding wheels. Essentially it consists of alumina, or aluminium oxide, the chemical symbol for which is Al_2O_3.

ALPHA IRON – The body centered cubic form of pure iron, stable below 1025°C.

ALUMEL – A nickel base alloy containing about 2.5% Mn, 2% AI, and 1 % Si, used chiefly as a component of pyrometric thermocouples.

ALUMINIUM – Grayish white metal, very light in weight, and having in its pure form low mechanical strength, frequently alloyed with other elements to improve its physical characteristics.

ALUMINIUM ALLOY – Aluminium which is alloyed with other metals to give it strength and desirable properties.

ALUMINIUM BRONZE – Alloy containing 90% copper and 10% aluminium, extensively used for diecasting.

ANTIFREEZE – A chemical added to the coolant in order to lower its freezing point.

ANTIFRICTION BEARINGS – Ball, roller and needle bearings exhibit very low friction and are suitable for very high speeds, and high loading.

ANTIMONY – Brittle, bluish white metallic element designated Sb. Melting point 630°C. Used as a constituent in some alloys, for instance, bearings and storage battery plates.

ARGON – An inert gas used in certain welding and heat treatment processes.

ARSENIC – A brittle, grayish metallic element designated As. Melting point 814°C. Used as a constituent in some alloys, and in the manufacture of lead shot.

ASBESTOS – A fibrous organic mineral that is non-combustible, non-conducting and acid resistant.

ATOM – The smallest particle of an element.

AUSTENITE – A solid solution of iron and carbon and sometimes other elements in which gamma iron, characterized by a face centered crystal structure, is the solvent. This is stable only within a particular range of composition and temperature, and is non-magnetic.

AUSTENITIC CAST IRON – Cast iron containing such a proportion of alloying constituents (nickel, chromium, copper or manganese) that the structure in the cast state is completely austenitic at ordinary temperatures.

BABBITT METAL – White metal bearing alloy, suitable for bearings subjected to moderate pressures, contains tin 59.5% min, copper 2.25-3.75%, antimony 9.5-11.5%, lead 26% min, iron 0.08% max, bismuth 0.08% max.

BACKING SAND – Foundry sand placed next to the facing sand after the latter is in place. It forms the bulk of sand used to complete the mould.

BAINITE – A structure in steel named after E.G. Bain that forms between 481° C and the M's temperature. At the higher temperatures, it is known as upper or feathery bainite. At the lower temperatures it is known as lower or a acicular bainite and resembles martensite.

BAKELITE – Trade name for one of the first used thermo-setting synthetic resins. It is derived from the name of the inventor Dr. L.H. Backeland, and its formation is the result of a chemical action between formaldehyde and phenol.

BAR – A piece of material thicker than sheet, long in proportions to its width or thickness, and whose width to thickness ratio is much smaller than sheet or plate, as low as unity for squares and rounds.

BARK – The decarborized layer just beneath the scale that results from heating steel in an oxidizing atmosphere.

BASE METAL – Metal present in the alloy in largest proportion.

BEARING METALS – Metals (alloys) used for that part of a bearing which is in contact with the journal *e.g.*, bronze or white metal, used on account of their low coefficient of friction when used with a steel shaft.

BELFAST SAND – Red moulding sand of fine grain, and good bonding qualities with moderate refractoriness, suitable for use as a facing sand.

BELL METAL – High tin bronze, used in the casting of bells, which is composed of up to 30% tin, together with some zinc and lead.

BESSEMER STEEL – Steel manufactured in a Bessemer converter, and sometimes referred to as mild steel.

BILLET – A solid semifinished round or square product that has been hot worked by forging, rolling or extrusion.

BLUE VITRIOL – A chemical mixture of copper sulphate, water and sulphuric acid. Applied to polished metal for layout purposes, it turns to copper colour.

BOND – In grinding wheels and other relatively rigid abrasive products, the material that holds the abrasive grains together. In welding, the junction of joined parts.

BORON CARBIDE – An abrasive used in cutting tools, a compound whose chemical formula is B_4C and obtained from borontrioxide (B_2O_3) and coke at a temperature of 2500°C. Fine powder as hard as diamond.

BRASS – A range of copper zinc alloys, usually those containing 55-80% copper. Alloys containing not less than 63% of copper are called ALPHA BRASSES. When less than 63% of copper is present, the alloy is called ALPHA-BETA alloy.

BRAZING ALLOY – Copper zinc alloy, which sometimes includes small percentages of tin, and lead, used for brazing, the melting point of which is governed by the percentage of zinc.

BRINE – Water that has been saturated or nearly saturated with salt.

BRIQUETS – Compact cylindrical or other shaped blocks formed of finely divided materials by incorporation of a binder, by pressure, or both. Materials may be ferroalloys, metal borings or chips, silicon carbide etc.

BRONZE – A copper rich, copper tin, copper lead or copper beryllium alloy to which often alloying elements (phosphorous, aluminium, zinc, silicon) may be added. Usually bronze is a copper tin alloy containing 90% copper and 10% tin.

BUILDING BRICK – These are made from clay. Generally, the clay is mixed with water to a plastic state and extruded in a column that is wire-cut crosswise to the desired size. Occasionally the dry pressing process is used.

CADMIUM – White ductile metallic element used to plate steel and as an alloying element.

CALCIUM ALUMINIUM SILICON – An alloy composed of 10-14% calcium, 8-12% aluminium, and 50-53% silicon used for degasifying and deoxidizing steel.

CALCIUM BORIDE – An alloy of calcium and boron, containing about 61% boron and 39% calcium and used in deoxidation and degasification of non-ferrous metals and alloys.

CALCIUM CARBIDE – A grayish black, hard crystalline substance made in the electric furnace by fusing lime and coke. Addition of water to calcium carbide forms acetylene and a residue of slaked lime.

CALCIUM MANGANESE SILICON – An alloy containing 17 to 19% calcium, 8 to 10% manganese, 55 to 60% silicon and 10 to 14% iron, used as a scavenger for oxides, gases and non-metallic impurities in steel.

CALCIUM MOLYBDATE – A crushed product containing 40-50% molybdenum, 23-25% lime, 3% iron (max) and 5-10% silica, used to add molybdenum to iron and steel produced in open hearth, air furnace or electric furnace.

CALCIUM SILICON – An alloy of calcium, silicon and iron containing 28-35% calcium, 60-65% silicon and 6% max iron used as a deoxidizer and degasifier for steel and cast iron. Sometimes called CALCIUM SILICIDE.

CAPPED STEEL – Semiskilled steel cast in a bottle top mould and covered with a cap fitting into the neck of the mould. The cap causes the top metal to solidify. Pressure is build up in the sealed in molten metal and results in a surface condition much like that of RIMMED STEEL.

CARBIDE – A compound of carbon with one or more metallic elements.

CARBOHYDRATES – Constitute a large group of molecules, widely distributed in nature, which contains only carbon, hydrogen and oxygen. The simplest carbohydrates are sugars.

CARBON – A non-metallic element found in all organic substances that is used as an alloying element in ferrous metals.

CARBON STEEL – Steel containing carbon up to about 2% and only residual quantities of other elements except those added for deoxidation, with silicon usually limited to 0.60% and manganese to about 1.65%. Also termed PLAIN CARBON STEEL.

CARBORUNDUM – Artificially manufactured abrasive, trade name for a carbide of silicon (SiC) which is prepared by heating sand with coke in an electric furnace.

CARTRIDGE BRASS – Alloy containing about 70% copper and 30% zinc, in which impurities are kept to a minimum, and it possesses a high degree of strength, combined with good ductility.

CAST ALLOY TOOL – A cutting tool made by casting a cobalt base alloy and used at machining speeds between those for high speed steels and sintered carbides.

CAST IRON – Iron obtained by slightly purifying the pig iron in a cupola or other furnace. This has high carbon content, averaging between 2.5 and 4.5% and frequently alloyed with small percentage of other elements and primarily used for making castings. It is somewhat brittle.

CELLULOSE – A polysaccharide of glucose units that constitutes the chief part of the cell walls of plants. For example, cotton fibre is over 90% cellulose and is the raw material of many manufactured goods such as paper, rayon and cellophane. In many plant cells, the cellulose wall is strengthened by the addition of lignin, forming lignocellulose.

CEMENT – Material used for uniting other materials so that they adhere permanently.

CEMENTED CARBIDE – A solid and coherent mass made by pressing and sintering a mixture of powders of one or more metallic carbides, and a much smaller amount of a metal, such as cobalt, to serve as a binder.

CEMENTITE – Hard, brittle, crystalline iron carbide (compound of iron and carbon Fe_3C) found in steels having a high carbon content. It is characterized by an orthorhombic crystal structure. When it occurs as a phase in steel, the chemical composition will be altered by the presence of manganese and other carbide forming elements.

CERAMIC – Metallic oxides of metals such as silicon and aluminium.

CERAMIC MATERIALS – The materials that demonstrate great hardness and resistance to heat and are used to make cutting tools, coatings on tools, parts subjected to very hot conditions, abrasives and mechanical parts.

CERMET (Ceramal) – A body consisting of ceramic particles bonded with a metal.

CESIUM 13T – A radioisotope, recovered as a fission product from nuclear reactors, with a half-life of 33 years and a dominant characteristic gamma radiation of 0.66 mev. It is suitable as a gamma radiation source, especially in radiography and therapy.

CHILL – (1) A metal insert embedded in the surface of a sand mould or core or placed in a mould cavity to increase cooling rate at that point. (2) White iron occurring on a gray iron casting such as the chill in the wedge test.

CHINESE SCRIPT – An angular microstructural form with the constituents alpha (AI-Fe-Si) and alpha (AI-Fe-Mn-Si) in cast aluminium alloys. A similar microstructure is found in cast magnesium alloys containing silicon as Mg_2Si.

CHROMEL – (1) 90% Ni, 10% Cr alloy used in thermocouples. (2) A series of Nickel chromium alloys, some with iron, used for heat resistant applications.

CHROMIUM – Grayish white metallic element obtained from chromite, chemical symbol is Cr and melting point 1830°C, used in alloying steels and corrosion resisting plating.

CLAD METAL – A composite material containing two or three layers that have been bonded together. The bonding may have been accomplished by rolling, welding, casting, heavy chemical deposition or heavy electroplating.

COAL TAR – Also called crude oil, when subjected to fractional distillation and purification, yields a variety of useful products-neutral, acidic, and base oils.

COBALT-60 – A radio isotope with a half-life of 5.2 years and dominant characteristic gamma radiation energies of 1.17 and 1.33 MeV. It is used as a gamma radiation source in industrial radiography and therapy.

COLD FINISHED STEEL – Steel bar which has been cold drawn/cold rolled, centerless ground or turned smooth to improve surface finish, accuracy or mechanical properties.

COLD ROLLED STEEL – Steel which has been passed through rollers at the steel mill to size it accurately and smoothly.

COLLOIDS – Finely divided material, less than 0.5 micron in size, gelatinous, highly absorbent and sticky when moistened.

COLUMNAR STRUCTURE – A coarse structure of parallel columns of grains having the long axis perpendicular to the casting surface.

COMBINED CARBON – The part of the total carbon in steel or cast iron that is present as other than FREE CARBON.

COMPOSITE FIBRES – The strands of material used as reinforcement extending through a resin or other matrix in a composite material. An example is carbon fibres in an epoxy matrix. Loads applied to the structure are carried by the fibres.

COMPOSITE MATERIAL – Materials exhibiting a much higher strength than the matrix or base material because of reinforcement fibres.

CONDUCTORS (electrical) – Materials in which an electromotive force causes appreciable drift of electrons, called CURRENT.

CONSTANTAN – A group of copper nickel alloys containing 45-60% copper with minor amounts of iron and manganese and characterized by relatively constant electrical resistivity irrespective of temperature used in resistors and thermocouples.

CONVERSION COATING – A coating consists of a compound of the surface metal produced by chemical or electro-chemical treatments of the metal.

COPPER – A reddish, soft, ductile metal with very good heat and electrical conductivity and is the basic element in brass and bronze.

CORE – (1) In a metal casting, the hollow parts (which cannot be shaped as easily by the pattern) that are made by using formed sand shapes, that are strengthened by baking or by using epoxy. (2) In a ferrous alloy, the inner portion that is softer than the outer portion or case.

CORE SAND – Variety of silica sand. Rock sand, river bed and sea shore sand, commonly known as sharp sand, used for making of cores in the foundry because they are capable of withstanding high temperatures, and resisting the penetrating action of the molten metal.

CORUNDUM – Natural abrasive of the aluminium oxide type that has higher purity than emery.

ROCUS CLOTH – A very fine abrasive polishing cloth.

CHROMIUM BRONZE – It is a precipitation hardening alloy of copper with upto 1 per cent chromium. It has high electrical conductivity and high temperature resistance.

CRUCIBLE – A vessel or pot, made of refractory substance or of a metal with a high melting point, used for melting metals or other substances.

CRUCIBLE STEEL – A high grade steel made by melting iron in a crucible and adding charcoal, pig iron and some substance rich in carbon so that the resulting metal will contain from 0.75-1.5% carbon. This steel is used for tools, dies and better grades of cutlery.

CRYSTAL – A solid composed of atoms, ions or molecules arranged in a pattern which is repetitive in three dimensions.

CRYSTALLOID – A substance that forms a true solution and is capable of being crystallized.

CUNIFE – Cunife is a copper-nickel iron alloy that is malleable, ductile and machinable, even in an age-hardened form. Magnets are formed from wire stock in round, square, or rectangular form.

CUPRO NICKEL ALLOY – Alloy of nickel and copper (approximately 60% nickel and 30% copper), which combines the strength of steel with immunity from corrosion and resistance to high temperature.

CURIE – The quantity of a radioactive nuclide in which the number of disintegrations per second is 3. 700 × 10 to the power of ten.

CUTTING FLUID – A fluid, usually a liquid, used in metal cutting to improve finish, tool life or dimensional accuracy.

DEGASIFIER – A material employed for removing gases from metals and alloys.

DELTA IRON – An allotropic (polymorphic) form of iron, stable above 1390°C, crystallizing in the body centered cubic lattice.

DENDRITE – A crystal that has tree like branching pattern, being most evident in cast metals slowly cooled through the solidification range. Dendrite generally grow inward from the surface of a mould.

DEOXIDIZER – A substance that is used to remove either free or combined oxygen from molten metals, for example, ferrosilicon in steel making.

DEVELOPER – (1) In photography, a processing solution that reduces the exposed grains of an emulsion to metallic silver, thus making the image visible. (2) In xero radiography a dry powder used to make the electrostatic image visible. (3) In penetrant inspection, a material used to draw the penetrant back to the surface, thus revealing locations of cracks or fissures.

DEVIZES SAND – Coarse greenish yellow moulding sand, suitable for dry sand work in the foundry for medium and heavy castings, but not for fine work, owing to its large grain size.

DIAMAGNETIC SUBSTANCES – Actually set up fields that oppose applied fields.

DIAMOND – Allotropic form (crystalline form) of carbon (the hardest known mineral) which when very strongly heated, changes to graphite. Used as a cutting tool, and a grinding tool and to dress grinding wheels.

DROSS – The scum that forms on the surface of molten metals largely because of oxidation but sometimes because of the rising of impurities to the surface.

DRY SAND MIXTURE (Mould) – Specially prepared and for making the moulds that are to be dried before using. This demands a sand that when dried or baked will give strength, porosity and permeability.

DUCTILE IRON – A high strength type of cast iron that will bend without fracturing.

DURALUMIN – Aluminium alloy containing copper, manganese and magnesium, which can be cast, forged or stamped, and is widely used for sheets, tubes, forgings, rivets, nuts, bolts and similar parts.

DYE PENETRANT – Penetrant with a dye added to make it more readily visible under normal lighting conditions.

ELASTOMER – Any of various elastic substances resembling rubber.

ELECTRIC STEEL – Special alloy steel, tool steel, and steel for castings, melted in electric furnaces that permit very close control and the addition of alloying elements directly into the furnaces.

ELECTRICAL INSULATING MATERIALS – The materials which offer a very large resistance to flow of current, and for that reason they are used to keep the current in its proper path along the conductor.

ELECTRICAL SHEETS – It is the trade name for iron and steel sheets used in the manufacture of punchings for laminated magnetic circuits and usually refers to silicon steel sheets.

ELECTROLYTE – A non-metallic conductor, usually a fluid, in which electric current is carried by the movement of ions.

ELECTROMAGNET – A magnet of variable strength produced by passing current through conductors around a soft iron core.

ELEKTRON – Magnesium base alloy supplied in the form of tubes, sheets, extruded sections, forgings and castings.

ELEMENT – A substance which cannot be chemically broken down to simpler substances.

EMERY – An abrasive material which, like corundum or aluminium oxide type, is a natural abrasive.

EMULSIFIER – (1) A material that increases the stability of dispersion of one liquid in another. (2) In penetrant inspection, a material that is added to some penetrants after the penetrant is applied to make a water washable mixture.

ENAMEL – Type of paint that dries to a smooth, glossy finish.

ERITH SAND – Yellow, close grained, refractory moulding sand, having good strength and reasonable permeability.

EUTECTIC – Mixture (an alloy) in which the proportions of the constituents are such that the mixture has a lower melting point than any of the constituents.

FACING SAND – Sand that forms the face of the mould which comes in contact with the molten metal.

FALKIRK SAND – Moulding sand with a coarse, open texture. It has very good permeability and moderate binding qualities.

FERRIC OXIDE – Red iron oxide, commonly available as haematite ore. Used in ground form in cores and moulds to increase hot compressive strength.

FERRITE – A solid solution of one or more elements in body centered cubic iron. Iron which contains little or no carbon. It is very soft and ductile and is known as alpha iron. A magnetic form of iron.

FERROALLOYS – Alloys containing of certain elements combined with iron, and used to increase the amount of such elements in ferrous metals and alloys. In some cases the ferroalloys may serve as deoxidizers.

FERROALUMINIUM – An alloy of iron and aluminium containing about 20% iron and 80% aluminium.

FERROCHROMIUM – An alloy of iron and chromium available in several grades containing from 60-72% chromium and from 0.06-7% carbon.

FERROMAGNETIC MATERIAL – A material that in general exhibits the phenomena of hysteresis and saturation, and whose permeability is dependant on the magnetizing force.

FERROMANGANESE – An alloy of iron and manganese containing from 78-82% manganese.

FERROMOLYBDENUM – An alloy of iron and molybdenum containing 58-64% molybdenum.

FERROPHOSPHOROUS – An alloy of iron and phosphorous containing 70% iron and 25% phosphorous.

FERROSILICON – An alloy of iron and silicon available in several grades containing different percentages of silicon from 14-20% silicon, 42-52% silicon, 69.5-82% silicon, 82-88% silicon and 88-95% silicon.

FERROUS – From the latin word FERRUM meaning iron, describes an alloy containing a significant amount of iron.

FERROUS METALS – All metals that are alloys of iron, carbon, and other materials.

FIBRE GLASS – A resin matrix reinforced with glass fibres for strength. A reinforced plastic manufacturing material with many applications.

FILTER – In radiography a device, usually, a thin metallic layer inserted into a beam of radiation so as to modify the transmitted spectrum of radiation. It may be used to enhance or reduce contrast or minimize undesirable scattered radiation.

FIRE BRICK – Brick made of refractory clay or other material which resists high temperatures.

FIRE CLAY – A type of clay which is resistant to high temperatures.

FIXER (hypo) – A photographic processing solution, the principle function of which is to dissolve the undeveloped silver halide grains from the developed film, thus making the image more prominent. It often serves also to harden the gelatine and halt the developing process.

FLUX – A solid, liquid or gaseous material that is applied to solid or molten metal in order to clean and remove oxides.

FOAM RUBBER – It is also called sponge. Foam rubbers are formed by the inclusion of chemicals in rubber compounding which form gases during vulcanization.

FREE CARBON – The part of the total carbon in steel or cast iron that is present in the elemental form as graphite or temper carbon.

FREE FERRITE – Ferrite that is structurally separate and distinct as may be formed without the simultaneous formation of carbide when cooling hypoeutectoid austenite into the critical temperature range.

GAMMA IRON – The face centered cubic form of pure iron, stable from 910-1230°C.

GANGUE – The worthless portion of an ore that is separated from the desired part before smelting is commenced.

GEL COAT – A thin coat of plastic resin covering fibreglass panels.

GILDING METAL – Alloy containing 80-90% copper, the reminder being zinc. Often used in wire form for jewellery and decorative applications.

GLACIER METAL – Tin base alloy used for lining bearings.

GLASS – Transparent substance produced by the fusion of sand and certain metallic salts, of which soda compounds are most common.

GLUE LAMINATED BEAM – A structural wood beam made by gluing thinner boards together until a desired dimension for beam thickness is reached. Glue laminated beam will support large loads and can span long distances with only end support.

GRAIN – Individual crystal in metals.

GRANITE – A rock composed of quartz, feldspar and mica from which dimensionally stable surface plates and angle plates are made.

GRANULAR PEARLITE – A structure formed from ordinary lamellar pearlite by long annealing at a temperature below but near to the critical point, causing the cementite to spherodize in a ferritic matrix.

GRAPHITE – Native carbon in hexagonal crystals, also foliated or granular massive, of black colour with metallic lusture, and soft.

GRAPHITE FIBRE – Strands of carbon in graphite form used in composite materials as the main load bearing constituent.

GRAPHITIZER – Any substance, such as silicon, titanium, aluminium etc. which promotes the formation of graphite in cast iron compositions.

GRAY CAST IRON – A cast iron that gives a gray fracture due to the presence of flake graphite. Often called GRAY IRON.

GRIT SIZE – Nominal size of abrasive particles in a grinding wheel corresponding to the number of openings per linear inch in a screen through which the particles can just pass. Sometimes called GRAIN SIZE.

GUN METAL – Bronze alloy containing 88-89% copper, 5-10% tin and 2-6% zinc, lead up to 20% may also be added, although the quantity seldom exceeds 5%.

HALF LIFE – The characteristic time required for half of the nuclei of a radioactive species to disintegrate spontaneously.

HALF VALUE LAYER – In radiation, the thickness of absorber that will reduce the intensity of radiation to one half. It is useful in estimating radiographic exposure.

HIGH CARBON STEEL – Steel that has more than 0.6% carbon.

HIGH SPEED STEEL – Alloy steel (alloying elements being tungsten, chromium, vanadium, cobalt and molybdenum) which retains its strength and hardness at red heat, and is thus suitable for cutting tools which reach high temperatures in use.

HINDU MINIUM – A high strength aluminium alloy containing, in addition to aluminium, magnesium, iron, titanium, copper, nickel and silicon, which after heat treatment has a strength exceeding that of mild steel.

HOT ROLLED STEEL – Steel rolled to shape while being heated to the plastic condition.

HOYT METAL – Commercial grade of white metal used for bearing purpose.

HYPER EUTECTIC ALLOY – Any binary alloy whose composition lies to the right of the EUTECTIC on an equilibrium diagram and which contains some eutectic structure.

HYPO EUTECTIC ALLOY – Any binary alloy whose composition lies to the left of the EUTECTIC on an equilibrium diagram and which contains some eutectic structure.

IMPURITIES – Elements or compounds whose presence in a material is undesired.

INCONEL – Nickel alloy highly resistant to heat and corrosion, with good mechanical properties, consisting of 80% nickel, 12-14% chromium, the balance being iron.

INERT GAS – A gas that may be used as a shield in welding or heat treatment to prevent oxidation or scaling.

INGOT – A large block of metal that is usually cast in a metal mould and forms the basic material for further rolling and processing.

INGOT IRON – Commercially pure open hearth iron.

INSULATING MATERIALS (electrical) – Materials which offer a very large resistance to flow of current and for that reason they are used to keep the current in its proper path along the conductor.

INOCULATED IRONS – Inoculated irons are high strength irons of such composition that they would ordinarily be white as cast are often inoculated in the ladle with a silicon compound to cause graphitization. Typical agents used are ferosilicon, calcium silicide, Si-Mn-Zr, or Ca-Mn-Si in crushed form.

INVAR – Nickel iron alloy (35-36% nickel, 0.5% carbon and 0.5% manganese, the remainder being iron) having a very low coefficient of thermal expansion at ordinary temperatures.

ION – An atom, or group of atoms, that has gained or lost one or more outer electrons and thus carries an electric charge. Positive ions, or cations, are deficient in outer electrons. Negative ions or anions, have an excess of outer electrons, thus ion is electrostatically charged.

IRIDIUM 192 – A radio isotope with a half-life of 74 days and 12 dominant characteristic gamma radiation energies ranging from 0.14-0.65 MeV. It is suitable as a gamma radiation source, mostly in radiography.

IRON – Silver white metallic element, symbol Fe, and melting point 1535°C. Pure iron consists of homogenous crystal grains generally referred to as ferrite.

IRON (wrought) – Malleable iron produced from molten pig iron by a working or puddling process which removes the impurities.

KANTHAL – It is an electrical resistance alloy of iron-chromium-aluminium with small additions of cobalt. About 25 per cent Cr, 5 per cent Al, 3 per cent Co, and balance almost pure iron.

KAOLIN – A fine white clay that is used in ceramics and refractories composed mostly of kaolinite, a hydrous silicate of aluminium. Impurities may cause various colours and tints.

KILLED STEEL – Steel that has been deoxidized with agents such as silicon or aluminium to reduce the oxygen content to such a level that no reaction occurs between carbon and oxygen during solidification. This prevents gases from evolving during solidification.

LACQUER – A quick drying automotive paint.

LAMELLAR – An alternating plate like structure in metals (as in pearlite).

LAMINATE – (1) A composite metal, usually in the form of sheet or bar, composed of two or more metal layers so bonded that the composite metal forms a structural member. (2) To form a metallic product of two or more bonded layers.

LAMINATIONS – Metal defects with separation or weakness generally aligned parallel to the worked surface of the metal.

LASER – Light Amplification by Simulated Emission of Radiation. A device in which heat is derived from the intense coherent beam of laser light energy. This intense, narrow beam of light is used in some welding and machining operations.

LEAD – Heavy, bluish grey, soft, ductile metal, which has a specific gravity of 11.3 and a melting point of 327°C, extensively used alone, and as the basis of many antifriction alloys.

LEAD SCREEN – In radiography, a screen is used (1) to filter out soft wave or scattered radiation and (2) to reduce the intensity of the remaining radiation so that the exposure time can be decreased.

LEDEBURITE – The eutectic of the iron carbon system, the constituents being austenite and cementite. The austenite decomposes into ferrite and cementite on cooling below the transformation temperature.

LIGNIN – A substance that is related to cellulose, that with cellulose forms the woody cell walls of plants and the material that cements them together. Methyl alcohol is derived from lignin in the destructive distillation of wood.

LOAM – Clayey sand mixture having the consistency of slime, and used in the making of moulds and cores for heavy castings.

LOW CARBON STEEL – Steel containing less than 0.3% carbon.

LUTE – Fine adhesive composition of substances such as clay, sharp sand, plumbago and horsedung tempered with water. Used for sealing joints in moulds and cores, for the purpose of making them air or metal tight.

MAGNESIUM – A very light metal (about 106 Ibs/cuft) that alloys readily with aluminium and other metals.

MAGNESIUM ALLOY – Alloy containing at least 85% of magnesium and having a specific gravity of 1.8, alloying elements include aluminium, manganese, zinc, and silicon. Widely used for aircraft components, their weight is only two thirds that of aluminium, and a quarter of that of steel.

MAGNETICALLY HARD ALLOY – A ferromagnetic alloy capable of being magnetized permanently because of its ability to retain induced magnetization and magnetic poles after the removal of externally applied fields, an alloy with high coercive force.

MAGNETICALLY SOFT ALLOY – A ferromagnetic alloy that becomes magnetized readily upon the application of a field and that return to practically a non-magnetic condition when the field is removed, an alloy with the properties of high magnetic permeability, low coercive force, and low magnetic hysterisis loss.

MAGNOLIA METAL – White metal bearing alloy containing 4.75-6% tin, 78-80% lead and 15-16% antimony.

MALLEABLE CAST IRON – A cast iron made by a prolonged anneal of WHITE CAST IRON in which decarbonization or graphitization, or both, takes place to eliminate some or all of the CEMENTITE. The graphite is in the form of temper carbon. This is less brittle than gray cast iron.

MANGANESE – A brittle, hard metallic element used as an alloy in steel to give it toughness to withstand wear and strain.

MANGANESE BRONZE – A group of special alloys, not really bronzes at all, containing about 1% manganese , 60% copper, 40% zinc and small traces of iron, tin, lead or aluminium, the total percentage of these not exceeding 5%.

MARTENSITE – An unstable constituent that is formed by heating and quenching steel. It is formed without diffusion and only below a certain temperature known as M's temperature. Martensite is the hardest of the transformation products of austenite, having an acicular or needle like microstructure.

MATTER – Any substance which occupies space and has weight. The three forms of matter are solids, liquids and gases.

MEDIUM CARBON STEEL – Steel with a carbon content of 0.3-0.6%.

METAL – An opaque lustrous elemental chemical substance that is a good conductor of heat and electricity and when polished , a good reflector of light.

METALLOID – A non-metal that exhibits some, but not all, of the properties of a metal. Examples are sulphur, silicon, carbon, phosphorous and arsenic.

METALLURGY – The science and study of the behaviours and properties of metals and their extraction from their ores.

MILD STEEL – Carbon steel with a maximum of about 0.25% carbon.

MOLECULE – The smallest portion to which a substance may be reduced by subdivision and still retain its chemical identity.

MOLYBDENUM – Element used in alloying steel, including high speed steel. It gives red hardness and increases the strength of steel at high temperatures. It increases the corrosion resistance of stainless steels at high temperatures, increases the machinability of carbon steels and reduces the temper brittleness of aluminium steels.

MONEL METAL – Trade name for a nickel copper alloy (67% nickel, 28% copper, 5% iron, manganese, and silicon combined) which exhibits high strength and toughness and corrosion resistance.

MU METAL – Special alloy of nickel and iron, also containing copper and manganese, requiring only a very small magnetizing force to produce a normal flux density *i.e.*, the alloy is said to have high permeability (Greek letter MU for permeability).

MUNTZ METAL – Alloy of brass family containing 60% copper and 40% zinc used for manufacturing condenser tubes.

MUSIC WIRE – A high carbon steel wire of the highest quality used for making mechanical springs.

MYCALEX – It is the trade name for a ceramic product made up of glass-bonded mica flakes that possess a combination of properties found in other insulating materials.

NATURAL RUBBER – Natural rubber is obtained in the form of a latex from the sap of Hevea brasiliensis and a few other plants. Crude rubber is coagulated by heat or by addition of electrolytes.

NAVAL BRASS – Alloy containing from 57.5-59.5% copper, 0.6-1.0% tin and not more than 0.75% of impurities, the balance being zinc (addition of tin improves the resistance of the alloy to corrosion by sea water). Used for under-water fittings of marine craft.

NEOPRENE – A synthetic rubber, highly resistant to oil, light, heat and oxidation.

NEUTRON – Elementary nuclear particle with a mass approximately the same as that of hydrogen atom and electrically neutral.

NICHROME – Alloy of nickel and chromium which is practically non-corrosive, can withstand high temperature without oxidation and is used for furnace components.

NICKEL – A strong, grayish, white, ductile metal, which has high resistance to oxidation and corrosion. Therefore, used in pure form for some applications, such as plating. It is more usually alloyed with other metals.

NICKEL BRONZE – Bronze alloy of which there are two main series (1) low nickel bronze (nickel below 5%) used, for bronze castings, and (2) high nickel bronze (nickel over 10%) resistant to heat, and to corrosive attack from chemical liquors.

NICKEL SILVER – Also called GERMAN SILVER. Alloy with composition copper 60%, zinc 20%, and nickel 20%. Class of alloys used in the manufacture of electrical resistance coils and elements, decorative articles for which its lustrous colour (which increases in whiteness with nickel content) make it very suitable, or for heavy duty works such as high pressure steam fittings.

NICROSILAL – A nickel-chromium alloy cast iron having a composition 1.7% carbon, 4.5% silicon, 0.8% manganese, 18.0% nickel, and 2% chromium, the balance is iron.

NIMONIC ALLOY – Nickel base alloy possessing high resistance to heat and corrosion, used for components in gas turbines and jet propulsion engines.

NIRESIST IRON – Alloy cast iron (typical composition 14% nickel, 1.5% silicon, 1 % manganese, and 3% carbon and remainder iron) which possesses exceptional resistance to heat and corrosion.

NISPAN ALLOY – Range of alloys having controlled expansion and elastic properties.

NITENSYL – Group of cast iron which have a tensile strength of 23-25 tons./sq.inch by suitable heat treatment. A typical composition is

1.5% nickel, 1.5% silicon, 2.9% carbon and 0.8% manganese, the balance being iron.

NITRALLOYS – Nitralloys are the steels developed for nitriding process. The commonly used grades contain 0.20 to 0.40 per cent carbon, 0.9 to 1.5 per cent Cr, 0.80 to 1.20 per cent Al, and small additions of Mo, Si, and Mn.

NODULAR CAST IRON – A cast iron that has been treated while molten with a master alloy containing an element such as magnesium or cerium to give primary graphite in the spherulitic form.

NODULAR GRAPHITE – Graphite or carbon in the form of spheroids.

NOMAG – Non-magnetic cast iron, used for castings in electric motors and alternators and similar applications. A typical composition is 11% nickel, 1.5% silicon, 3% total carbon, up to 7% manganese, the balance being iron.

NON-FERROUS – Metals and alloys which do not contain any large proportion of iron, examples being brass, copper, aluminium and lead.

NUCLEUS – (1) The first structurally stable particle capable of initiating recrystallization of a phase or the growth of a new phase and possessing an interface with the parent matrix. (2) The heavy central core of an atom in which most of the mass and the total positive electric charge are concentrated.

NYLON – A group of plastics of nitrogenous structure known as polyamides which are crystalline in nature and can be so processed as to orient the crystals axially thus making the tensile strength of fibres extremely high.

OIL STONE – An abrasive stone that is oiled and used to sharpen cutting tools.

ORANGE PEEL – A pebble grained surface which develops in forming of metals having coarse grains.

ORE – A natural mineral that may be mined and treated for the extraction of any of its components, metallic or otherwise.

OSMIUM – Osmium is the heaviest of all metals (sp gr. 22. 48), which melts at 4900°F and is harder than glass and quartz.

PARAMAGNETIC MATERIALS – These materials are only feebly magnetic.

PARTING SAND – Fine sand used for dusting on sand mould surfaces that are to be separated.

PEARLITE – The laminar mixture of ferrite and cementite in slowly cooled iron carbon alloys as found in steel and cast iron.

PEARLITIC MALLEABLE IRON – Irons made from the same or similar chemical compositions as regular malleable iron, but so alloyed or heat treated that some of the carbon in the resultant material is in the combined form.

PERMANENT MAGNET – Special magnet steel that retains its magnetic power indefinitely.

PETROCHEMICALS – Chemicals derived from petroleum substances or materials manufactured from a component of crude oil or natural gas.

PEWTER – Alloy containing 1.8% lead, 89.4% tin, 7% antimony and 1.8% copper.

PHASE – It is a portion of matter which is homogeneous in the sence that its smallest adjacent parts are indistinguishable from one another.

PHASE DIAGRAM – Phase diagram is also called equilibrium diagram or constitution diagram, indicates the relative amount and composition of phases present in an alloy at a given temperature and pressure, when the alloy is in equilibrium.

PHOSPHOROUS – One of the elements, its chemical symbol is P. Its formula weight is 123.92, specific gravity 1.82, and melting point 44.1°C.

PHOSPHOR BRONZE – Alloy containing 78.5-81.5% copper, 9-11 % tin, 9-11 % lead, 0.05-0.25% phosphorous and 0.75% zinc, has excellent antifriction properties. Used as bearing material.

PHOTON – The smallest possible quantity of an electromagnetic radiation that can be characterized by a definite frequency.

PIG IRON – Iron produced from iron ore in the blast furnace, basic raw material from which all cast iron, wrought iron and steel are made. Usually contains about 4.5% carbon and impurities such as phosphorous, silicon and sulphur.

PITCH – Usually coal tar pitch obtained in the manufacture of coke and distilled off at about 175°C.

PLASMA – An ionized gas of extremely high temperature achieved by passing an inert gas through an electric arc. Plasma arcs are used in welding, cutting and machining processes.

PLASTIC – A certain group of natural and synthetic resins and their compounds that can be moulded, cast, extruded or used for coatings and films.

PLASTIC ELASTOMERS – Plastic elastomers are materials which exhibit the characteristics of rubber, but are of a basic chemical structure that is decidedly different from that of natural rubber.

PLATINUM – It is a silver-white heavy metal, unaffected by acids, air, or a great variety of chemical agents. It is extensively used, either solid or clad, for chemical equipment.

POLYESTERS – Polyesters are a reaction product of polyhydric alcohol and a dibasic acid plus monomer styrene or diallyl phthalate. In combination with glass fibres they form a product which has an outstanding strength-weight ratio.

POLYETHYLENES – The product of straight chain polymerization of ethylene and are obtainable as viscous liquids, gums, and tough flexible solids suitable for moulding.

POLYMER – A chemical compound or mixture of compounds formed by polymerization and consisting essentially of repeating structural units.

PORCELAIN – Porcelain is a ceramic product made up of clays, quartz, and feldspar used as high voltage insulator.

POWDER METALLURGY – Forming parts out of powdered metal by compacting the powder into a mould under great pressure and heating it.

PRECIOUS METAL – One of the relatively scarce and valuable metals– gold, silver and platinum group of metals.

PROTON – The positively charged particle in the nucleus of an atom.

PRUSSIAN BLUE – A blue pigment, obtainable in tubes which is used to find hot spots in a bearing.

QUICK SILVER – Metallic mercury.

RADIO ACTIVE ELEMENT – An element which has at least one isotope that undergoes spontaneous nuclear disintegration to emit positive alpha particles, negative beta particles or gamma rays.

RADIO ISOTOPE – An isotope that emits ionizing radiation during its spontaneous decay.

RADIUM – A radio active element. It is found in nature as radium 226, which has a half-life of 1620 years.

RAZOR STEEL – Steel containing 1.15-1.25% carbon. This steel is forged at 816°C, and hardened at 750-775°C. It is tempered at 230°C to straw colour.

RED BRASS – A brass containing approximately 85% copper, 5% zinc, 5% tin and 5% lead.

RED ROCK SAND – Open grain moulding sand obtained from red sand stone rocks, with good permeability but low bonding qualities.

REFRACTORY – Materials that will resist change of shape, weight, or physical properties at high temperatures say exceeding 1000°C. These materials are usually silica, fire clay, diaspore, alumina and kaolin. They are used for furnace linings.

RESIDUAL ELEMENTS – Elements present in an alloy in small quantities but not added intentionally.

RESIDUE – The material that remains after completion of a chemical or physical process, such as combustion, distillation, evaporation or filtration.

RESISTORS – Poor conductors.

ROUGHING STONE (hone) – A coarse honing stone.

RIMMED STEEL – A low carbon steel (insufficiently deoxized) that during solidification releases considerable quantities of gases (mainly carbon monoxide). When the mould top is not capped, a side and bottom rim of several centimeters forms. The solidified ingot has got scattered blow holes and porosity in the center but a relatively thick skin free from blow holes.

RUST – A corrosion product containing hydrated oxide of iron. Applied only to ferrous alloys.

SAND (moulding) – Substance used in foundries for making the moulds.

SCRAP – Materials or metals that have lost their usefulness and are collected for reprocessing.

SEALANT – A sealing agent that has some adhesive qualities, it is used to prevent leakage.

SEMICONDUCTORS – A few substances containing metallic elements, have considerably less electrical conductivity. These contain a few electrons to give them conductivity intermediate between metals and insulators.

SEMISTEEL – Cast iron to which a small proportion of mild steel or wrought iron scrap is added during the melting of the pig iron so that the product will have a lower carbon content than the average iron, from 2.5-3.2% and which is tougher.

SILICA – Silicon dioxide, SiO_2 occurring in nature as quartz, opal etc.

SILICON – Non-metallic element which can be added to steel, cast iron and non-ferrous alloys. It acts as a DEOXIDIZER, and also tends to form graphite by throwing the carbon out of solution and thereby increases the impact resistance of the steel, and, up to a silicon content of 1.75%, the elastic limit is increased also.

SILICON ALUMINIUM – An alloy of 50% silicon, and 50% aluminium used for making silicon additions to aluminium alloys.

SILICON BRASS – A series of alloys containing 0.5-0.6% silicon, 1-19% zinc and a substantial amount of copper.

SILICON BRONZE – Alloy containing about 90-95% of copper, to which is added silicon and manganese, equivalent in strength to medium carbon steel with resistance to corrosion and fatigue.

SILICON CARBIDE – A refractory and abrasive material made by sand, coke, and saw dust in an electric arc furnance.

SILICON CARBIDE BRIQUETS – Silicon carbide in BRIQUET form used as an inoculant and deoxidizer in cupola melted gray iron.

SILICON COPPER – An alloy of silicon and copper, used as a deoxidizer and hardener in copper base alloys.

SILMANAL – It is the name given to a rather expensive alloy of silver, manganese, and aluminium that has unusual magnetic properties for special applications.

SILVER – A white, ductile metal that is an excellent conductor of heat and electricity.

SILVERY IRON – A type of pig iron containing 8-14% silicon, 1.5% carbon max, 0.06% sulphur max and 0.15% phosphorous max.

SLAG – The more or less completely fused and vitrified matter separated during the reduction of a metal from its ore.

SLURRY – A watery mixture of insoluble material such as mud, lime or plaster of paris.

SMOG – The irritating haze resulting from the sun's effect on certain pollutants in the air, notably those from automobile exhaust. Also a mixture of fog and smoke.

SMOKE – Solid or liquid particles under 1 micron in diameter. Particles suspended in air after incomplete combustion of materials containing carbon. The matter in the exhaust emission which obscures the transmission of light.

SODIUM SILICATE – Na_2SiO_3-Also called water gas.

SOLDERING ALLOY – Fusible alloy used to join together two metallic surfaces with the aid of heat. Soft solder is an alloy of lead and tin, in which the proportions of the two constituents may vary from almost pure lead to almost pure tin.

SOLDERING FLUID – Liquid flux used when soldering.

SOLID SOLUTIONS – Solid solutions are alloys containing alloying elements that are relatively soluble in the base metal in the solid state.

SOLUBLE OIL – Specially prepared oil whose water emulsion is used as a cutting or grinding fluid.

SOLUTE – A substance that is dissolved in a solution and is present in minor amounts.

SOLVENT – A substance that is capable of dissolving another substance and is the major constituent in a solution.

SORBITE – Structure consisting of evenly distributed carbide of iron particles in a mass of ferrite, formed when a fully hardened steel is tempered at between 550 and 650°C.

SPELTER – Hard solder used during brazing containing 60% copper, 20% tin and 20% zinc.

SPHEROIDITE – It is the structure in steel, in which cementite takes the form of rounded particles, or spheroids, instead of plates.

STAINLESS STEEL – Steel which resists corrosion by the atmosphere and the attack of acids and which does not scale when subjected to high temperature. Alloy steels containing iron, atleast 11 % chromium, nickel, molybdenum and 0.1-1 % carbon.

STEATITE – Steatite is the name given to a fired ceramic product which contains 80 per cent or more talc bonded with ceramic fluxes to a non-porous structure.

STEEL – An alloy of iron and less than 2% carbon plus some impurities and small amounts of alloying elements is known as plain carbon steel. The alloy steels contain substantial amounts of alloying elements such as chromium or nickel besides carbon.

STELLITE – Non-ferrous alloy containing 35-80% cobalt, 10-40% chromium, 0-25% tungsten and 0-10% molybdenum.

SYNTHETIC MATERIALS – A complex chemical compound which is artificially formed by the combination of two or more compounds or elements.

TERNARY ALLOY – An alloy that contains three principal elements.

THERMIT – Powdered form of finely divided iron oxide and aluminium which burns intensely to produce superheated liquid steel at a temperature of about 30.35°C, used for welding wrought iron and steel forgings and castings.

THERMOPLASTICS – Materials which when heated begin to soften at temperatures as low as 56.5°C, then can be moulded without any change in chemical structure.

THERMOSETTING MATERIALS – Materials that undergo a chemical change when moulded and cannot be resoftend by heating to reshape them.

TIN – A silvery white, soft metal used in solders and as a plating material.

TITANIUM – A strong, grayish metal that weighs less than steel.

TOOL STEEL – A special group of steels that is designed to specific uses, such as heat resistant steels that can be heat treated to produce certain properties mainly hardness and wear resistance.

TROOSTITE – Structure in steel (consisting of very finely divided iron carbide in what is known as "alpha iron") produced either by tempering a martensitic steel at between 250° and 450°C or by quenching steel at a speed sufficient to suppress the thermal change point fully.

TUNGSTEN – Hard, grayish, semiprecious metal with very high melting point of 3300°C, used for electrical contacts, filaments in electric lamps etc. Used as an alloying element in high speed steel.

TUNGSTEN CARBIDE – An iron gray powder composed of carbon and tungsten and used in sintered form as a cutting tool material.

TUNGUM BRONZE – Trade name for high strength bronze.

TWIN CRYSTAL – A crystal grain in which the crystal lattices of two parts are related to each other in orientation as mirror images across the interface known as the twinning plane.

VANADIUM – A rare metal used as an alloying element in steel to improve shock resistance and forgeability.

VULCANATES – Vulcanates are materials which reduce plasticity of the rubber compound, while maintaining its elasticity.

WHITE IRON – An extremely hard cast iron that results from pouring the hot metal into a mould with a chill plate in it.

WROUGHT IRON – Contains 1-2% slag, which is distributed through the iron as threads and fibres imparting a tough fibrous structure. Usually contains less than 0.1 % carbon. It is tough, malleable, and relatively soft.

WROUGHT METALS – These are metals furnished in the shapes resulting from the operations such as rolling, forging, drawing and extrusion.

YELLOW BRASS – An alloy of about 70% copper and 30% zinc.

ZINC – Bluish, grey metal with a melting point of 418°C, it becomes brittle at 200°C and can be powdered at this temperature.

ZIRCON – Natural zirconium silicate, containing when pure 67.3% zirconium oxide, and 32.7% silica, and is used as a moulding medium.

CHAPTER 3

MATERIAL PROPERTIES

ACICULAR STRUCTURE – A microstructure characterized by needle shaped constituents.

ALLOTROPY – Ability of a material to exist in several crystalline forms.

AMORPHOUS – Non-crystalline, a random orientation of the atomic structure.

ANISTROPY – A material that has specific physical properties in different directions. Rolled steel is strongest in the direction of rolling.

API GRAVITY – Gravity expressed in units of standard American Petroleum Institute (hydrometer).

AUSTENITE – A solid solution of cementite or iron carbide, Fe_3C in iron.

BANDED STRUCTURE – A segregated structure of nearly parallel bands aligned in the direction of working.

BEL – A unit denoting the ratio of power levels of signals or sound. The number of bels may be given as the common logarithm of the ratio of powers.

BETA RAY – A ray of electrons emitted during the spontaneous disintegration of certain atomic nuclei.

BOILING POINT – The temperature at which a liquid begins to boil.

BOUND ELECTRONS – The inner orbit of electrons around the nucleus of the atom.

BREAKING POINT – The final rupture of a material which is being pulled in tension, after it has reached its ultimate strength.

BRINELL HARDNESS – The hardness of metal or alloy measured by pressing a hard ball (usually 10 mm diameter) with a standard load into the specimen. A number is derived by measuring the indentation with a special microscope.

BRITTLE METAL – A metal which exhibits only a very small change in dimensions before it fractures.

BRITTLENESS – The property of materials to not deform under load, but to break suddenly, for example, cast iron and glass are brittle. Brittleness is opposite to plasticity.

BULK MODULUS OF ELASTICITY – Ratio of a uniform, triaxial (equal in all directions) tensile or compressive stress to the change in volume it produces.

CAVITATION – The formation and instantaneous collapse of innumerable tiny voids or cavities within a liquid subjected to rapid and intense pressure changes.

CEMENTATION – Process of introducing elements into the outer layer of metal objects by means of high temperature diffusion.

CEMENTITE – Iron carbide, Fe_3C, a hard brittle, crystalline compound observed in the microstructure of iron base alloys.

CHAFING FATIGUE – Fatigue initiated in a surface damaged by rubbing against another body.

CHLORINATION – A refining or degasification process, wherein dry chlorine gas is passed through molten aluminium base and magnesium base alloys to remove entrapped oxides and dissolved gases.

CLEAVAGE – Splitting (fracture) of a crystal in a crystallographic plane of low index.

CLEAVAGE FRACTURE – A fracture, usually of a polycrystalline metal, in which most of the grains have failed by cleavage, resulting in bright reflecting facets. It is one type of crystalline fracture.

CLEAVAGE PLANE – A characteristic crystallographic plane or set of planes on which cleavage fracture easily occurs.

COALESCENCE – The union of particles of a dispersed phase into larger units, usually effected at temperatures below fusion point.

COHESIVE STRENGTH – (1) The hypothetical stress in an unnotched bar causing tensile fracture without plastic deformation. (2) The stress corresponding to the forces between atoms.

COLD SHORT – A condition of brittleness existing in some metals at temperatures below the recrystallization temperature.

COLD SHUT – (1) A discontinuity that appears on the surface of cast metal as a result of two streams of liquid meeting and failing to unite. (2) A portion of the surface of a forging that is separated, in part, from the main body of metal by oxide.

COLUMNAR STRUCTURE – A coarse structure of parallel columns of grains having the long axis perpendicular to the casting surface.

COMPLETE FUSION – Fusion which has occured over the entire base metal surfaces exposed for welding.

COMPOUND – A combination of two or more elements that are mixed together.

COMPRESSIBILITY – The property of a substance (*e.g.*, air) by virtue of which its density increases with increase in pressure.

COMPRESSIVE STRENGTH (ultimate) – The maximum stress that can be applied to a brittle material in compression without fracture.

COMPRESSIVE STRENGTH (yield) – The maximum stress that can be applied to a metal in compression without permanent deformation.

COMPRESSIVE STRESS – Compressive stress is compression load per unit area perpendicular to the load.

CONDUCTIVITY – The quality or power of conducting or transmitting heat, electricity etc.

CONGRUENT TRANSFORMATION – An isothermal or isobaric phase change in which both of the phases concerned have the same composition throughout the process.

COOLING STRESSES – Residual stresses resulting from nonuniform distribution of temperature during cooling.

CORROSION – The destructive chemical or electro-chemical reaction of a material and its environment, usually associated only with metals in contact with liquids.

CORROSION EMBRITTLEMENT – The severe loss of ductility of a metal resulting from corrosive attack, usually intergranular and often not visually apparent.

CORROSION FATIGUE – Effect of the application of repeated or fluctuating stresses in a corrosive environment characterized by shorter life than would be encountred as a result of either the repeated or fluctuating stresses alone or the corrosive environment alone.

COUPON – A piece of metal from which a test specimen is to be prepared, often an extra piece as on a casting or forging.

COVALENT BOND – A bond between two or more atoms resulting from the completion of shells by the sharing of electrons.

CRAZING – Minute surface cracks on the surface of materials often caused by thermal shock.

CREEP – Slow plastic deformation in steel and most structural metals caused by prolonged stress under the yield point at elevated temperatures.

CREEP LIMIT – (1) The maximum stress that will cause less than a specified quantity of creep in a given time. (2) The maximum nominal stress under which the creep strain rate decreases continuously with the time under constant load and at constant temperature. Sometimes called CREEP STRENGTH.

CRITICAL POINT – The temperature or pressure at which a change in crystal structure, phase, or physical properties occur.

CRYSTALLIZATION – Act or process of forming crystals or bodies formed by elements or compounds solidifying so that they are bounded by plane surfaces.

CRYSTAL UNIT STRUCTURE OR UNIT CELL – The simplest polyhedron that embodies all the structural characteristics of a crystal and makes up the lattice of a crystal by indefinite repetition.

CURIE TEMPERATURE – The temperature of magnetic transformation below which a metal or alloy is magnetic and above which it is paramagnetic.

DAMAGING STRESS – The minimum stress which, if exceeded in the material, would render the part unfit for service before the end of its normal expected life.

DAMPING CAPACITY – The ability to absorb vibration. More accurately defined as the amount of work dissipated into heat by a unit volume of material during a completely reverse cycle of unit stress.

DENDRITE – A crystal formed during solidification of a metal or alloy characterized by a structure like that of a fir tree.

DENSITY – The ratio of the mass of a body to its volume.

DUCTILE or MALLEABLE METAL – A metal that may be worked to a different size or shape without breaking or shattering.

DUCTILITY – The property of a material to deform permanently or to exhibit plasticity, elongation or bending or twisting without rupture (breaking or cracking) while under tension.

DYNAMIC CREEP – Creep that occurs under the conditions of fluctuating load or fluctuating temperature.

ELASTIC DEFORMATION – The movement or deflection of a material when an external load is applied that is less than the elastic limit.

ELASTICITY – The ability of a material to return to its original form after the load has been removed.

ELASTIC LIMIT – Maximum stress that can be applied to a metal without causing plastic deformation that will remain after the load is relaxed to zero.

ELASTIC RATIO – It is the ratio of yield point stress to tensile strength of a metal.

ELECTROLYSIS – Chemical change resulting from the passage of an electric current through an electrolyte.

ELONGATION – Elongation is the strain produced by uniaxial tension.

EMBRITTLEMENT – Reduction in the normal ductility of a metal due to a physical or chemical change.

ENDURANCE LIMIT – The limiting stress below which the metal will withstand without fracture an infinitely large number of cycles of stress.

ENDURANCE RATIO – It is the ratio between the endurance limit to tensile strength.

EUTECTIC – The alloy which has the lowest melting point possible for a given composition.

EUTECTOID – A solid solution of any series which cools without change to its temperature of final composition.

FACTOR OF SAFETY – The ratio of the damaging stress to working stress.

FATIGUE IN METALS – The tendency in a metal to fail, breaking or cracking under conditions of repeated cyclical stressing that take place well below the ultimate tensile strength.

FATIGUE STRENGTH – The amount of stress that can be applied to a metal without failure while it is subjected to ten million or more cycles of load reversals. In mild steel, the fatigue strength is about 50 per cent of the tensile strength.

FERRITE – Iron practically carbon free. It forms a body centered cube lattice and may hold in solution considerable amounts of silicon, nickel or phosphorous.

FLUIDITY – Ability of molten metal to flow readily, usually measured by the length of a standard spiral casting.

FRACTURE STRESS – The maximum principal true stress at fracture. Usually refers to unnotched tensile specimens.

FREE ELECTRONS – Electrons which are in the outer orbit of the atoms nucleus.

GRAIN BOUNDARY – The outer perimeter of a single grain where it is in contact with adjacent grains.

GRAIN GROWTH OR RECRYSTALLIZATION – Metal grains begin to reform to larger and more regular size and shape at certain temperatures, depending to some extent on the amount of prior cold working.

GRAIN REFINER – A material added to a molten metal to attain finer grains in the final structure.

GRAIN SIZE – For metals, a measure of the area or volume of grains in polycrystalline material, usually expressed as an average when the individual sizes are fairly uniform. Reported in terms of number of grains per unit area or volume, average diameter, or as a grain size number derived from area measurements.

GRANULAR PEARLITE – A structure formed from ordinary lamellar pearlite by long annealing at a temperature below but near to the critical point, causing the cementite to spheroidize in a ferrite matrix.

GROWTH – With reference to cast iron, permanent increase in volume that results from continued or repeated cyclic heating and cooling at elevated temperatures.

HALF-LIFE – The characteristic time required for half of the nuclei of a radioactive species to disintegrate spontaneously.

HARDENABILITY – The property that determines the depth and distribution of hardness in a ferrous alloy induced by heating and quenching.

HARDNESS – The property of metal to resist being permanently deformed. This is divided into three categories, resistance to penetration, resistance to abrasion and elastic hardness.

HOT SHORT – Brittleness in hot metal. The presence of excess amounts of sulphur in steel causes hot shortness.

HYDROGEN EMBRITTLEMENT – A condition of low ductility in metals resulting from the absorption of hydrogen.

IMPACT TEST – A test in which small notched specimens are broken in Izod-Charpy machine. This test determines the notch toughness of a metal.

INCLUSIONS – Impurities, usually oxides, sulphides, silicates and similar compounds, retained from the ore or from processing.

INFRARED RADIATION – Electromagnetic energy with wave lengths from 770 to 1200 nanometers.

ISOTROPY – Quality of having identical properties in all directions.

KISH – Graphite thrown out by liquid cast iron in cooling.

KNOOP HARDNESS – Microhardness determined from the resistance of metal to indentation by a pyramidal diamond indenter, having edge angles of 172° and 32 minutes and 130°, making a rhombohedral impression with one long and one short diagonal.

LATTICE, SPACE – The term that is used to denote a regular array of points in space. For example, the sites of atoms in a crystal. The points of the three dimensional space lattice are constructed by the repeated application of the basic translations that carry a unit cell into its neighbour.

LIQUIDUS – The temperature at which freezing begins during cooling and ends during heating under equilibrium conditions, represented by a line on a two phase diagram.

MACHINABILITY – The relative ease of machining that is related to the hardness of the material to be cut.

MACROSTRUCTURE – The structure of metals as revealed by examination of the etched surface of a polished specimen at a magnification not exceeding ten diameters.

MALLEABILITY – In metals, the property of being able to undergo mechanical deformation (flattened by rolling or hammering) without rupturing or developing a marked increase in resistance to change of shape.

MELTING POINT – The temperature at which a pure metal, compound or eutectic changes from solid to liquid, the temperature at which the liquid and the solid are in equilibrium.

MICROPOROSITY – Porosity visible only with the aid of microscope.

MICROSTRUCTURE – The structure of polished or etched metal specimens as seen enlarged through a microscope.

MODULUS OF ELASTICITY – A measure of the rigidity of metal. Ratio of stress, within proportional limit, to corresponding strain. Also called ELASTIC MODULUS and COEFFICIENT OF ELASTICITY and YOUNGS MODULUS.

MODULUS OF RIGIDITY – Ratio of shearing stress to shearing strain, within the proportional limit.

MODULUS OF RUPTURE – A fictitious stress calculated for either bending or torsion tests on the basis of the load causing failure, and the assumption that the elastic equation for stress applies up to the point of failure.

Ms TEMPERATURE – The temperature at which martensite begins to form in an alloy system on cooling.

MOTTLED – White iron structure interpersed with spots or flecks of gray.

NOTCH BRITTLENESS – Susceptibility of a material to brittle fracture at points of stress concentration.

NOTCH DUCTILITY – The percentage reduction in area after complete separation of the metal in a notch tensile test.

NOTCH RUPTURE STRENGTH – The ratio of applied load to original area of the minimum cross-section in a stress rupture test of a notched specimen.

NOTCH SENSITIVITY – A measure of the reduction in strength of a metal caused by the presence of stress concentration.

NOTCH TOUGHNESS – The resistance to fracture of a metal specimen having a notch or groove when subjected to a sudden load, usually tested on an Izod-Charpy testing machine.

OILINESS – The capacity of the lubricant to stick on to the surface under conditions of heavy load.

OXIDATION – The slow or rapid reaction of oxygen with other elements, burning. In metals, over oxidation during heating under oxidizing conditions often results in permanent damage to metals.

OXIDATION REDUCTION – A chemical reaction in which one or more electrons are transferred from one atom or molecule to another.

PEARLITE – A micro constituent of iron and steel consisting of alternative layers of ferrite and iron carbide or cementite.

PEEL LOAD – In metal, plastics or composites, the force that acts to peel apart joined pieces.

PH – The negative logarithm of the hydrogen ion activity. It denotes the degree of acidity or basicity of a solution.

PERMEABILITY – In casting of metals, the term is used to define the porosity of foundry sands in moulds and the ability of trapped gases to escape through the sand.

PERMANENT SET – When a metal remains deformed from its original dimensions after the forces applied to it have been reduced to zero, it is said to have undergone plastic deformation, and the amount of deformation is called the permanent set.

PHASE – A portion of an alloy, physically homogeneous throughout, that is separated from the rest of the alloy by distinct BOUNDARY surfaces. The following phases occur in the iron carbon alloy, molten alloy; austenite, ferrite, cementite, and graphite.

PHYSICAL CHANGE – A change which does not alter the composition of the molecules of a substance.

PIEZOELECTRIC EFFECT – The reversible interaction, exhibited by some crystalline materials between an elastic strain and an electric field. The direction of the strain is dependant upon the polarity of the field.

PLASTICITY – The quality of material such that it can be deformed without breaking. Clay is completely a plastic material. Metals exhibit plasticity in varying amounts.

PLASTIC DEFORMATION – Deformation that does or will remain permanent after the removal of load which caused it.

POISSON'S RATIO – When a rod of elastic material is elongated by stretching (strain), the lateral (crosswise) dimensions will contract. Poisson's ratio is the ratio between the strain and the amount of lateral contraction.

PROPORTIONAL LIMIT – Proportional limit is the value of stress at which the stress strain curve first bends to the right. The proportional limit is high for steels and low for cast iron, copper, and aluminium.

PROOF LOAD – A predetermined load, generally some multiple of the service load, to which a specimen or structure is submitted before acceptance for use.

PROOF, RESILIENCE – Maximum amount of elastic energy that may be stored without permanent deformation in the material during its first loading cycle.

RADIOACTIVITY – The spontaneous nuclear disintegration with emission of corpuscular or electromagnetic radiation.

RESIDUAL STRESS – Stress induced within the structure of a material by cold working, machining, and heat treatments.

RESILIENCE – Ability of a material to store elastic energy without permanent deformation.

RETENTIVITY – The capacity of a material to retain a portion of the magnetic field set up in it after the magnetizing force is removed.

ROCKWELL HARDNESS – Hardness test that uses a penetrator and known weights. Several scales are used to cover very soft to very hard materials. The Rockwell C scale is used mostly for steel.

SACRIFICIAL ANODE – A metal slug, usually magnesium and zinc, designed to concentrate galvanic corrosion upon itself and thus save a more important structure on which the anode is attached, such as a ship's hull or a burried pipe line.

SCRATCH HARDNESS – The hardness of a metal determined by the width of a scratch made by a cutting point drawn across the surface under a given pressure.

SEGREGATION – Non-uniform distribution of alloying elements, impurities or microphases.

SHEARING – A concentration of forces in which the bending moment is virtually zero and the metal tends to tear or to be cut along a transversal axis at the point of applied pressure.

SHEAR LOAD – A load that tends to force materials apart by application of side slip action.

SHEAR STRESS – Shear stress is the shear load per unit area, exerted between the material on one side of an imaginary plane of separation in the part and the material on the other side.

SHORTNESS – A form of brittleness in metal. It is designated as cold, hot and red to indicate the temperature range in which the brittleness occurs.

SKELP – A name of the semifinished steel of which butt-welded pipe is made.

SLAG (dross) – A fused product that occurs in the melting of metals and is composed of oxidized impurities of a metal and a fluxing substance such as limestone. The slag protects the metal from oxidation by the atmosphere since it floats on the surface of the molten metal.

SLIP PLANES – Also called slip bands. These are lines that appear on the polished surface of a plastically deformed metal. The slip bands are the result of crystal displacement, defining planes in which shear has taken place.

SOLID SOLUTION – Found in metals at temperatures below the solidus. Some of the types of solid solutions are continuous, intermediate, interstitial, substitutional and terminal.

SOLIDUS – Seen as a line in a two phase diagram, it represents the temperatures at which freezing ends when cooling, or melting begins when heating under equilibrium conditions.

SOLUBILITY – The degree to which one substance will dissolve in another.

SPALLING – Breaking small pieces from a surface, often caused by thermal shock.

SPECIFIC GRAVITY – A numerical value that represents the weight of a given substance with the weight of an equal volume of water. The specific gravity for pure water is taken as 1.000.

SPHEROIDIZING – Consists of holding carbon steel for a period of time at just under the transformation temperature (heating and cooling). An aggregate of globular carbide is formed from other microstructures such as pearlite.

SPRING BACK – The tendency of a formed metal part to return to some extent to its former shape because of the elasticity of the metal.

SPUTTERING – To dislodge atoms from the surface of a material by collision with high energy particles for the purpose of depositing a metallic film on a part.

STIFFNESS – The ability of a metal or shape to resist elastic deformation. For identical shapes, the stiffness is proportional to the modulus of elasticity.

STRAIN -- The measure of the change in the size or shape of a body, referred to its original size or shape. This is unit deformation of a metal when stress is applied.

STRENGTH – The ability of a material to resist external forces. It is called tensile, compressive, or shear strength depending on the load.

STRESS – The load per unit area on a stress strain diagram.

STRESS, TENSILE – Refers to an object loaded in tension, denoting the longitudinal force that causes the fibres of a material to elongate.

STRESS, COMPRESSIVE – Refers to a member loaded in compression, which either gives rise to a given reduction in volume or a transverse displacement of the material.

STRESS, SHEAR – Refers to a force that lies in a parallel plane. The force tends to cause the plane of the area involved to slide on the adjacent planes.

STRESS, TORSION – The shearing stress that occurs at any point in a body as the result of an applied torque or torsional load.

STRESS RAISER – Can be a notch, nick, weld under cut, sharp change in section, or machining grooves or hairline cracks that provide a concentration of stresses when the metal is under tensile stress. Stress raisers pose a particular problem and can cause early failure in members that are subjected to many cycles of stress reversals.

STRESS RELIEF ANNEAL – The reduction of residual stress in a metal part by heating it to a given temperature and holding it there for a suitable length of time. This treatment is used to relieve stresses caused by welding, cold working, machinning, casting and quenching.

SUPER-CONDUCTIVITY – Property of zero electrical resistance (and zero magnetic induction) acquired by some metals at very low temperatures.

SURFACE TENSION – Interfacial tension between two phases one of which is a gas.

TARNISHING – A chemical reaction of a metal with its environment which results in formation of films which coat the metal.

TEMPER – (1) In ferrous metals, the stress relief of steels that are hardened by quenching for the purpose of toughening them and reducing their

BRITTLENESS. (2) In non-ferrous metals, temper is a condition produced by mechanical treatment such as cold working. An alloy may be cold worked to the hard temper, fully softened to the anealed temper, or two intermediate tempers.

TENSION LOAD – A load applied to joined parts that attempts to separate them by a pulling or stretching action.

TENSILE STRESS – Tensile stress is the tension load per unit area perpendicular to the load.

THERMAL CONDUCTIVITY – The quantity of heat that is transmitted per unit time, per unit cross-section, per unit temperature gradient through a given substance. All materials are in some measure conductors of heat.

THERMAL EXPANSION – The increase of the dimension of a material that results from the increased movement of atoms caused by increased temperature.

THERMAL SHOCK – A stress induced on the surface of a material such as carbide tools or fire brick caused by a rapid rate of heating and surface expansion.

THERMAL STRESS – Shear stress that is induced in a material due to unequal heating or cooling rates. The difference of expansion and contraction between the interior and exterior surfaces of a metal that is being heated or cooled is an example.

THERMOPLASTIC – Material capable of softening or fusing when heated and of hardening again when cooled.

THERMOSETTING – Material capable of becoming permanently rigid when cured by heating, will not soften by reheating.

TOUGHNESS – Ability of a metal to absorb energy and deform plastically before fracturing. Work per unit volume required to fracture a metal and is equal to the area under the stress strain curve.

TOXICITY – The degree of danger posed by a substance to animal or plant life.

TRANSFORMATION TEMPERATURE – The temperature at which one phase transforms into another phase, for example, where ferrite or alpha iron transforms into austenite or gamma iron.

TRANSITION TEMPERATURE – The temperature at which normally ductile metals become brittle.

ULTIMATE STRENGTH – Maximum conventional stress, tensile, compressive or shear that a material can withstand. The highest strength that a metal exhibits after it begins to deform plastically under load. Rupture of the material occurs either at the peak of its ultimate strength or at a point of further elongation and at a drop in stress load.

VALENCY – The capacity of an atom to combine with other atoms to form a molecule. The inert gases have zero valence, valence is determined by considering the positive and negative atoms as determined by the atoms gaining or losing of valence electrons.

VISCOSITY – The property of the fluids, either liquid or gaseous, which causes them to resist instantaneous change of shape or instantaneous rearrangement of their parts, due to internal friction.

VOID – A cavity or hole in a substance.

WEAR – A surface deterioration of contacting surfaces that destroys their operating relationship, or causes rupture if carried far enough.

WELDMENT – A unit formed by welding together an assembly of pieces.

WORKING STRESS – The actual stress that the part will be designed to handle.

YIELD POINT – The stress at which a marked increase in deformation occurs without an increase in load stress as seen in mild steel or medium carbon steel. This phenomenon is not seen in non-ferrous metals and other alloy steels.

CHAPTER 4

PROCESSES

AGING OF A PERMANENT MAGNET – It is the process of normal or accelerated change, under continued normal or specified artificial conditions, in the strength of the magnetic field maintained.

ARC WELDING – Method of welding or uniting two metallic pieces in which the metal is melted by the heat of an electric arc.

ATOMIC HYDROGEN WELDING – Welding of metallic pieces in which heat is liberated by hydrogen atoms when combining into molecules, is used to fuse the metal.

AUTOMATIC ARC WELDING – Method of arc welding in which the arc moves along the joint to be welded, feeds the electrodes to the arc, and governs the arc length, by automatic means.

BEADING – Process of forming a bead or lapped edge on a sheet metal article.

BENDING (by forging) – In bending there is a thinning of the material, accompanied by a spreading of the metal on the inside of the bend and a narrowing at the outside.

BLANKING – Cutting or shearing a shape (called blank) with a die from sheet metal stock. The hole material is saved and used for further operation.

BLAST CLEANING – Blast cleaning involves the forcing of a stream or spray of sand or other abrasive material against the surface of metal, stone, and other materials by means of compressed air.

BORING – Opening out or increasing the diameter of an existing drilled or cored hole by means of a boring tool.

BRAZING – Joining two pieces of metal without melting either one by using a brazing alloy (copper zinc alloy *i.e.*, brass) that melts at a lower temperature than the materials being joined.

BROACHING – Consecutive shearing of a hole or contour by a series of stepped cutting edges similar to a saw used in low acting presses for accurate sizing of holes or contours, such as gear teeth, and keyways.

BURNISHING – Bright, polished finish produced on the surface of a metal by rubbing it with another metallic harder surface, which smooths out small scratch marks.

BUTT-WELDING – Form of electrical resistance welding, the passage of current between the ends of the sections to be joined causing a rise in temperature sufficient to fuse the metal.

CALENDERING – A process that involves rolling of the product into sheets to achieve the desired surface finishes and thickness.

CASTING – Process of producing a metal object by pouring molten metal into a mould.

COINING (embossing) – Shaping a piece of a metal in a mould or die often creating raised figures or numbers.

COLD DRAWING – Reducing the cross section of a metal bar or rod by drawing it through a die, at a temperature below the recrystallization range, usually room temperature.

COLD ROLLING – Reducing the cross-section of a metal bar in a rolling mill below the recrystallization temperature, usually room temperature.

COLD SAWING – Any sawing process in which the chips are not heated to the softened state.

COLD WORKING – Deforming a metal plastically at a temperature below its lowest recrystallization temperature. Strain hardening occurs as a result of this permanent deformation.

CONFINED FLOW – Confined flow is the basis of drop forging and hot pressing. In this, the metal is ultimately confined in all directions, being forced to behave as pasty fluid in filling every portion of the confining cavity.

CENTERLESS GRINDING – Method of grinding metallic parts in which the piece to be ground (circular piece) is supported on a work rest, and passed between a grinding wheel running at a high speed and a controlling wheel running at a slow speed.

CENTRIFUGING – Casting of molten metals by using centrifugal force instead of gravity. The mould (or moulds) is rotated about a centre where molten metal is poured and allowed to follow sprues outward and get into the mould cavity.

CENTRIFUGAL CASTINGS – Castings of cylindrical design are made by the introduction of molten metal into revolving permanent moulds, known as centrifugal casting machines, where the centrifugal force is employed to bring pressure in filling of the mould.

CLADDING – The joining of one metal (usually sheet or plate) to another by using heat and pressure or by an explosive force. With this method, a thin sheet of more expensive metal or one less likely to corrode may be applied to a less expensive metal or one more likely to corrode.

CLIMB MILLING – Milling process in which the work is fed in the same direction as the path of the teeth on the cutter, below the arbor.

COUNTER SINKING – It is the opening out of the ends of a hole to form a conical hollow for receiving the head of a countersunk screw, rivet or bolt.

CRIMPING – Producing flutes or corrugations. Often used to gather metal as for stovepipe joints.

CROWNING – Shaping of the rim of a belt pulley so that the diameter at the centre of the face is greater than at the edges so as to keep the belt on the crown of the pulley.

CUPPING – Process in which a flat blank is converted into a cup like form.

CUT OFF – An operation that shears a stamping from a strip or bar.

DEFORMATION – Alteration of the form or shape as a result of the plastic behaviour of a metal under stress.

DIECASTING – Casting metal into a metallic mould by using pressure instead of gravity or centrifugal force.

DIFFUSION – The process of atoms or other particles intermingling within a solution. In solids, it is a slow movement of atoms from areas of high concentration towards areas of low concentration. The process may be (*a*) migration of interstitial atoms such as carbon, (*b*) movement of vacancies or (*c*) direct exchange of atoms to neighbouring sites.

DINKING – Cutting of non-metallic articles from the sheet, usually involving such operations as blanking and piercing.

DRAW FILING – Finishing operation in filing during which the file is moved in the direction of the greater length of the work, being held like a spokeshave.

DRAWING – Process, in press work, which involves reducing the diameter or cross-sectional dimensions of a cup, shell, tube, bar or wire.

DRILLING – Power or hand operated method for the production of holes in metal or other solid materials with a drill.

DROP FORGING – Drop forging is the operation in which a metal part is formed by repeated hammer blows on a bar or billet placed between a pair of dies containing the impression of the finished shape desired.

EDGING – Edging is a gathering operation. Here the metal is displaced to the desired shape by striking it between two dies. As the dies strike, the stock, metal is gathered toward the center of the cavity and some sideways movement also takes place.

ELECTRON BEAM WELDING – The fusion of material by energy imparted from an intense beam of electrons.

ELECTROPLATING – Coating an object with a thin layer of metal through electrolytic deposition.

ELECTRIC RESISTANCE WELDING – Uniting the parts by heating them to welding temperatures and then forcing the ends together by mechanical pressure.

EMBOSSING – Operation of raising a design or form above the surface of a component by means of high pressure effected by pressing or squeezing action.

ETCHING – Process of marking a metal by eating into it with an acid or other chemical.

EXFOLIATION – Cracking of the outer skin of the metal.

EXTRUSION – Process in which metal (often heated) is caused to flow through a restricted orifice by using an extremely high force, so creating an extremely elongated strip of uniform, but comparatively small cross-section.

FELLOWS PROCESS – Method of generating involute gear teeth by the use of pinion shaped cutter.

FETTLING – Operation of removing any sand left on the casting from the mould and core, and also the removal of surplus metal that is always associated with castings direct from the foundry, such as runners, risers, feeders and so on.

FILAMENT WINDING – A composite manufacturing process where the end product is to have a hollow internal shape. A filament of the fibre is wound around a form, then bonded in place with the resin matrix.

FILE CUTTING – Incising the teeth on a file blank by means of a mechanically driven chisel of high speed steel.

FILING – Operation of smoothing a rough surface or reducing the thickness of a piece of material with a file.

FITTING – Finishing of mating parts to dimensions which will allow the desired tightness or freedom of movement on assembly.

FLANGING – any process producing a flange. A flange may drawn on a shell or a tube.

FLASH WELDING – Method of electric resistance welding of sections of material by the formation of an electric arc between the edges of the pieces to be joined and then pressing together with a light pressure the molten edges.

FLATTING – Finishing operation carried out at the end of the work cycle to remove the various hammer marks on the surface left in by the previous shaping operations.

FLOAT GLASS – A glass manufacturing process that produces a continuous sheet or ribbon of glass.

FORGING – A method of metal working in which the metal is hammered into the desired shape, or is forced into a mould by pressure or hammering, usually after being heated to a more plastic state. Hot forging requires less force to form a shape than that of cold forging, which is usually done at room temperature.

FORM GRINDING – Grinding of tool designed for machining and other operations, in such a way that they are provided with the precise form required for their work, or regrinding them to restore the form after it has been lost as a result of service.

FORMING – Production of shaped part either by means of feeding in a tool ground to shape or form of the part or by what is known as spinning.

FOUNDRY – Place where metal is melted and poured into required shapes.

FULL AUTOMATIC – Process in which all phases, once started, are accomplished without the need of further manual input.

FULLERING – Similar to drawing and is a preliminary forging operation which results in an elongated section between two heavier sections.

FUSION – Merging of two materials while in a molten state.

FUSION SAWING – Sawing material in the cold state by means of friction discs, where a quick and rough cut is sufficient for the purpose.

FUSION WELDING – Welding process in which the metals are brought to the temperature at which they melt, and are joined without hammering.

GALVANIZING – The application of a layer of zinc to the surface of iron and steel for protection from corrosion.

GAS WELDING – Method of fusion welding in which a flame produced by the combustion of gases is employed to melt the metal.

GEAR CUTTING – Production of gear wheels by the various forms of shaping tools or rotary cutters.

GEAR HOBBING – Method of generating gear teeth by the use of a rotating worm shaped cutter.

GEAR PLANING – Production of gear wheels by the use of cutters having a sliding instead of a rotary action.

GENERATING GEAR TEETH – Production of correctly formed gear teeth automatically, without the use of cutter of intricate shape.

GRAVITY DIE CASTING – Production of casting by pouring molten metal into metallic moulds under the force of gravity only.

GRINDING – Finishing operation designed to give to parts already machined the necessary precision of form and accuracy of dimensions by the removal of excess material due to the cutting action of grains of abrasive in a wheel or disc.

HAMMER WELDING – Uniting by heating two pieces of mild steel or wrought iron to a soft malleable condition and hammering them together on an anvil.

HEADING – A metal gathering or upsetting operation. Originally used for production of screw and rivet heads in cold working process.

HONING – Process whereby a mirror finish is given to important bearing surfaces by using hones or abrasive tools that normally employ both rotary and longitudinal motion.

HOOKER PROCESS – It is usually a cold extrusion process and is commonly employed for the production of small, thin walled copper and aluminium seamless tubes and small cartridge cases.

HOT PRESSING – Forming or forging tough metals such as alloy steel at high temperatures.

HOT PRESSURE WELDING – In this, bonding of two materials is brought about by application of high pressure and production of mutual deformation at temperatures below the melting range of either. Bonding results from interlocking of the atoms of each piece with some additional interlocking introduced by diffusion.

HOT ROLLING – A process of forming metals between rolls in which the metals are heated to temperatures above the transformation range.

HYDROJET – A manufacturing process in which a material is cut by a high pressure jet of water often containing an abrasive material to enhance cutting action.

IMPACT EXTRUSION – In this process, a slug of metal is placed on a solid bottom die, and the impact of the punch causes the metal to flow back over the punch, which has a uniform section, slightly relieved.

INDEXING – Rotation of a workpiece by small uniform amounts, with or without the aid of change wheels, mainly in connection with milling operation.

INTERNAL GRINDING – Mechanical grinding of the internal bores of gears, bushes and wide variety of machine parts and articles of specialized character.

JOGGLING – Bending operation necessary for making a lap joint, yet keeping the top surface flush.

KNURLING – A cold working process in which a series of sharp serrations on a hardened steel roller are pressed into the material being knurled.

LANCING – A special form of piercing in which the entire contour is not cut, the blanked material remaining as a tab.

LAPPING – Finishing process following grinding, and designed to produce an exceptionally high degree of surface finish as well as a perfectly true surface accurate to size within extremely close limits, by using very fine abrasives.

LIMITS OF ACCURACY – Under a limit system, the workman aims at producing a component, the acceptable size of which may fall between two limits of size.

MAAG PROCESS – The process combines precision grinding of the flanks of the gear teeth with a generating action which ensures that a correct involute form will be maintained, or restored if distortion has taken place as a result of hardening and heat treatment.

MACHINING – Machining is the art of using machine tools and cutting tools in combination to reduce a piece of material to some specified shape and dimensions.

METALLIZING – Process of spraying metal on to a metallic or non-metallic surface, for building up of worn parts and for the protection of components against corrosion.

MILLING – Operation in which a workpiece is given a specific shape or form by means of a rotating cutter having many cutting teeth.

MILL WRIGHTING – General maintenance, repair and the making of parts for machinery in a factory. A millwright is a craftsman in the true sense of the word.

MOULDING – Practise of making moulds in which to pour molten metal to form castings.

NOTCHING – Operation of cutting gaps on the edge of an article.

OPEN DIE FORGING or SMITH FORGING – A forging process in which a drop hammer delivers blows of great force to a heated metal that is shaped by manipulating it under the hammer.

OXY ACETYLENE CUTTING – A method of cutting heavy iron or steel plates with the aid of an oxyacetylene torch.

PEENING – Work hardening the surface of metal by hammering or blasting with shot (small steel balls). Peening introduces compressive stresses on weld surfaces that tend to counteract unwanted tensile stresses.

PERFORATING – Piercing many small holes close together.

PICKLING – Process of cleaning (that is removing oxide film from) castings by dipping them into an acid bath prior to plating, painting or further cold working.

PIERCING – Producing holes in a blank by driving a punching tool into the mass of the metal, while at the same time forcing the displaced material into the particular form required by the part, this is known as deep piercing.

PINNING – Scratching of the work surface during a filing operation caused by small particles of metal getting wedged in front of the teeth of the file.

PLANING – Producing flat or plane surface on a moving or other part by removing metal from them by means of a cutting tool used in a planing machine.

PLANISHING – Hammering operation by which the surface of a metal component is brought to a first class condition for such operations as polishing, plating and spraying.

PLATE MOULDING – Specialized form of moulding which consists of one or more patterns mounted on a plate with the runner and ingates being allowed for in the pattern.

PLATE EDGE PLANING – Smoothing and generally making true and accurate the edges of large plates.

PLATING – The process of depositing a layer of one metal on another, often done electrically, for the purpose of corrosion protection, appearance, improved electrical conductivity, and other engineering requirements.

POWDER METALLURGY – Forming parts out of powdered metal by compacting the powder into a mould under great pressure and heating it.

PRESSING (hot) – Process of shaping a metal article by pressure and while the metal is hot, and without any interstage forging.

PRESSURE WELDING – Method of uniting two sheets of metal by squeezing them between heated dies.

PROFILING – Method of milling irregular forms by reproducing the form of a master template, form or pattern.

PULTRUSION – A process that is opposite to extrusion and is used in composite part manufacturing.

PUNCHING – The operation of cutting a hole in sheet metal using a die. The hole material is scrapped. It is a shearing operation carried out in a press.

PUSH FIT – Class of fit which allows two parts to be assembled under hand pressure.

REFINING OF METAL – Process of removal of impurities from metals or alloys, resulting in improved properties or a different product. Also refers to the refining of the grain of a coarse grained metal or alloy.

REFLEX PROCESS – Photo printing process enabling copies to be made from opaque originals.

RESISTANCE WELDING – Uniting two pieces of metal by the passage of a heavy electrical current (high amperage current) while the surfaces are pressed together.

RIVETING – A hammering operation in which the end of a metal pin (*i.e.*, a rod or rivet) is pressed over or spread out. It may be either a hot or cold working process.

ROLL FORMING – Process of passing strip stock between driven rollers that form the strip to the shape desired.

ROLL GRINDING – The operation of grinding large rolls of steel mills or calendars.

ROLL THREADING – A method of threading a part by pressing and rolling between serrated dies.

RUBBER METAL BONDING – Bonding of natural rubber or synthetic rubber to metal surfaces.

RUNNING FIT – Term used of parts which are assembled so that they are free to rotate a type of clearance fit.

SAND BLASTING – Process used for cleaning metal surfaces, consists of directing a stream of air under pressure, into which quartz sand is introduced as an abrasive, on to the work.

SCREW CUTTING – Process of cutting a screw thread, usually in a lathe, which involves copying a master screw called the leading or lead screw in the nature of its pitch, but not its profile.

SEAMING – A bending and flattening of an interlocking fold, *e.g.*, a stove pipe seam.

SEAM WELDING – Process of closing a seam by a continuous resistance weld.

SEASONING – Process by which the internal stresses existing in a piece of material (which have been subjected to sudden changes of temperature, as in casting or hardening or have been acted upon by heavy forces in some machining operations) are removed.

SEMI AUTOMATIC – A process in manufacturing that requires some degree of manual input, but acts without this input for at least part of the cycle.

SERRATING – Method of forming by the use of a knurling tool a series of straight serrations on the outer surface of a part, and parallel with it along the axis of a part.

SHAPING – Operations performed on shaping machines.

SHAVING – A finishing operation that cuts a small amount of material from the edge of a stamping to gain finish, accuracy and / or a square edge.

SHEARING – Cutting in a line by two opposed blades somewhat in a manner of the ordinary household shears.

SHELL MOULDING – A form of gravity casting process metal (usually a high melting temperature metal) in which the mould is made of a thin shell of refractory material.

SHIELDED ARC WELDING – Process in which the molten weld metal is protected from deterioration by an envelope of chemically reducing or inert gas such as helium or argon.

SHRINKAGE FIT – Method of fitting a shaft or other part into an undersized hole by heating the outer member until has expanded sufficiently.

SHOT PEENING – A cold working process in which the surface of a finished part is pelted with finely ground steel shots or glass beads to form a compression layer.

SILVER BRAZING – Brazing similar or dissimilar materials by using an alloy of silver or other metals.

SINTERING – The process of fusing compacted material such as metal powders into a solid or porous piece by applying heat sufficient to bond, but not melt, the particles.

SLOTTING – Machine tool operation designed for finishing slots or other enclosed parts not capable of being machined by normal planing or shaping machine.

SLUSH DIE CASTING – Casting process which involves the filling of a metallic mould with liquid metal or alloy, then inverting the mould and pouring out the unfrozen metal from the center.

SMELTING – The process of heating ores to a high temperature in the presence of a reducing agent such as carbon (coke) and of a fluxing agent to remove the gangue.

SOLDERING – Operation of joining two or more parts together by molten metal.

SPINNING – Cold drawing ductile sheet metal blanks into cylinders and other shapes having rotational symmetry in a spinning lathe.

SPOT WELDING – Method of uniting sheet material by a series of localized welds, produced by overlapping the edges of two sheets of metal and fusing them together between copper electrode tips at suitably spaced intervals, by means of a heavy electrical current. It is a form of resistance welding.

STAMPING – Process of shaping metal under a falling weight.

STUD WELDING – Method of attachment of studs, screws, pins and similar parts to plates and components by projection welding or arc welding.

SUNDERLAND PROCESS – Method of generating gears by the use of a rack shaped cutter.

SURFACE GRINDING – Method of grinding designed to carryout the removal of metal from the surface of a part or parts less expensively, and with greater precision than could be achieved by machining processes with cutting tools of steel or by hand or machine filing.

SURFACING – Movement of a lathe tool at right angles to the bed, to work on the face or end surface of the work.

SWAGING – Compacting or necking down metal bars or tubes by hammering or rotary forming.

SWEDGING – A cold die forging operation in which the metal is confined and made to flow plastically into the punch and/or die impressions.

TAPER FIT – Type of fit in which a taper on the mating surfaces is combined with an interference fit between the parts when assembled.

TAPER TURNING – Method of turning a taper on a work in the lathe.

TAP GRINDING – Sharpening of a tapping tool by grinding a taper on the end and a clearance behind the cutting edge along this taper.

TAPPING – Operation of producing an internal thread by means of a tap.

TEMPER ROLLING – Process used in the production of tin -plate, in which the annealed coils or strips are subjected to a small amount of cold reduction in a four high mill in order to increase the stiffness of the dead soft material without unduly reducing its ductility.

THERMIT WELDING – Method of uniting iron or steel parts by surrounding the joint with thermit mixture at a sufficiently high temperature to fuse the adjacent surfaces of the parts together.

THREAD GRINDING – Grinding of screw threads so as to remedy such main variations in form or size as may have occurred as a result of the hardening treatment of a part or tool.

THREAD MILLING – Operation of producing threads, both external and internal, by means of thread milling cutters, either single or multi-ribbed, according to the type of thread required and the design of the thread milling machine employed.

THREAD ROLLING – Method of producing threads on screws, bolts, screw caps etc. by rolling under pressure so as to make contact with the rollers, to which the required pitch and form of screw threads have previously been given.

TONGUE SHAPING – Forming a narrow projection (called a tongue) from solid metal on a shaping machine.

TREPANNING – Removal of a circular piece of material from inside a steel bar, plate or billet.

TRIMMING – Cutting away excess material left from previous operations.

TUMBLING – Process of cleaning, polishing, or crushing in which the articles to be treated are mixed with balls, pieces of hard material or abrasive and rotated in a more or less horizontal container.

TURNING – Process of reducing the diameter of materials held in a lathe and the general name given to the process whereby material which is turned by means of a driven spindle to which it is attached, is brought into contact with a stationary tool having cutting edges.

UPSETTING – Process of increasing the cross-sectional dimensions when forging, with consequent reduction in length. May be done manually or by machine.

UPSET FORGING – The process of increasing the cross-section of stock at the expense of its length.

VULCANIZATION – The process of treating crude or synthetic rubber or similar plastic material chemically to give it useful properties, such as elasticity, strength and stability.

WELDING – Joining of metals by the application of heat, without the use of solder or any other metal or alloy having a lower melting point than the metals being joined.

WHEEL TRUING – Cutting off irregularities on a rotating grinding wheel with a diamond dresser.

WORM GRINDING – Grinding the threads of worm gear wheels as a means of finishing them after the hardening operation.

WIRE DRAWING – Reduction in diameter of metal rods by drawing them through conical openings in the blocks.

CHAPTER 5

MACHINES AND DEVICES

ACCELERATING PUMP – A small cylinder and piston fitted to some types of SI engine carburettor, and connected to the throttle so as to provide a momentarily enriched mixture when the engine is accelerated.

ACCUMULATOR – A device used for storing liquid under pressure (sometimes used to smooth out pressure surges in a hydraulic system).

ACTUATOR – A device which uses fluid power to produce mechanical force and motion.

AFTER COOLER – A device used on the turbocharged engines to cool the air which has undergone compression.

AIR COMPRESSOR – A device used to increase air pressure.

AIR COOLED ENGINE – An engine that is cooled by the passage of air around the cylinders, not by passage of a liquid through water jackets.

AIR CONDITIONER – A device used to control the temperature, humidity, cleanliness and movement of air inside a room or a car.

AIR HAMMER – Power hammer, used for roughening out heavy forgings in foundry work.

ALTERNATOR – A electrical generator that produces alternating current (flow of current is considered to change in direction with each half cycle).

ARBOR PRESS – A hand operated machine capable of applying high pressure for the purpose of pressing parts together or removing parts.

AUTOMATIC LATHE – Specialized development from capstan and turret lathes, with a full automatic cycle of turning, boring and drilling of the workpiece, the function of the operator consists of loading or taking finished work from the machine.

AUTOMATIC SCREW MACHINE – Fully automatic single spindle or multiple spindle bar stock turret lathe.

AUTOMATIC WELDING MACHINE – Machine designed to carryout high speed electric arc or resistance welding in mass production processes, combining rapid production with accurate control of temperature.

BABCOCK AND WILCOX MILL – Dry grinding mill using rotary steel balls.

BALANCE BOX – A box, filled with heavy material used to counter balance the weight of the job and load of a crane of the cantilever type.

BALANCE CRANE – A crane with two arms, one having counterpoise arrangements to balance the load taken by the other.

BALANCING MACHINE – A machine for testing the extent to which a revolving part is out of balance, and to determine the weight and position of the masses to be added or removed, to obtain balance.

BAND SAW – A narrow endless strip of saw blading running over and driven by pulleys, as a belt used for cutting wood or metal to intricate shapes.

BAR LATHE – A small lathe of which the bed consists of a single bar of circular, triangular or rectangular section.

BATTERY – An electrochemical device for storing energy in chemical form so that it can be released as electricity. It is a group of electric cells connected together.

BED PLATE – A cast iron or fabricated steel base, to which the frame of an engine or other machine is attached.

BENCH LATHE – A lathe of small dimensions that can be mounted on a bench or stand.

BENDING MACHINE – Machine designed to bend and fold sheet metal.

BLOWER – A low pressure air pump, usually of one rotary or centrifugal type.

BOARD DROP STAMP – A stamping machine in which the frictional grip of opposed rollers on either side of a vertical board lifts a tup, which falls when the roller pressure is released.

BOLT MAKING MACHINE – A machine which forges bolt by forming a head on a round bar.

BOOST FAN – A fan for restoring the pressure drop of air or gas, used for restoring the pressure drop in transmission pipes, and for supplying air to furnaces.

BORING AND TURNING MILL (vertical) – Machine designed for boring and turning castings and forgings.

BORING MACHINE (horizontal) – Machine used for boring, the spindle being horizontal. In one type, the spindle only rotates and in another type the spindle rotates and also has a horizontal movement.

BORING MACHINE (vertical) – Machine used for boring, the spindle being vertical, very similar to a radial driller. Also called BORING MILL.

BRASS FINISHERS LATHE – Lathe specially designed with attachments to machine brass work in quantities. The chief feature is the provision of special hand operated rests.

BREAK LATHE – Heavy lathe with sliding bed to accommodate large work. The machine comprises a fast and a loose head stock, and a base plate upon which the bed is mounted.

BROACHING MACHINE – Machine designed to drive a tappered tool of special form, known as a broach, through a hole or over a piece of work, which bring the hole or the surface to the desired finished size.

CAPSTAN LATHE – Lathe designed to use a number of cutting tools mounted on a rotating turret or capstan, and arranged to perform turning operations successively.

CENTRE LATHE – Machine for carrying out turning, boring, screw cutting operations on a work held between centers or in a chuck, but not for repetition work.

CENTRIFUGAL GOVERNOR – A governor which uses fly weight force to sense speed in order to control the fuel supplied to the combustion chambers.

CENTRIFUGAL PUMP – A pump using the centrifugal force produced by a rapidly rotating impeller to displace liquid.

CENTRIFUGE – A device with a rapidly rotating bowl which separates the impurities of a fluid by intense centrifugal force.

CIRCUIT BREAKER (lighting system) – A device that opens the circuit when the current draw becomes excessive and closes when the current flow is reduced.

CIRCULATING PUMP – The term applied to cooling water/ lubricating oil pumps which effect circulation of fluid.

COMPRESSOR – A mechanical device to pump air, and thereby increase the pressure.

CONDENSER (electrical) – An arrangement of insulated conductors and dielectrics for the accumulation of an electric charge.

CUPOLA – Special type of blast furnace, chiefly used for the melting of cast iron.

CYLINDRICAL GRINDER – Grinding machine designed to true up and bring to size cylindrical parts such as shafts, spindles, rollers etc.

DIESEL ENGINE – An internal combustion engine having fuel injected into the combustion chamber near the end of the compression stroke. The fuel is ignited by the heat of compression only.

DIODE – A device that allows current to pass but only in one direction.

DISC GRINDER – Grinding machine embodying one or more abrasive discs mounted horizontally on a vertical spindle, the work reposing on the surface of a flat rotating carrier or table.

DRAWCUT SHAPER – Type of machine with a traverse tool movement. In this, the work is fixed to the table which reciprocates under the tool, the tool moving across the table for the purpose of obtaining the feed.

DRILL GRINDER – Machine employed for the accurate grinding of the cutting points and lips of twist drills.

DRILL PRESS – A fixed machine to drive a tool in rotary motion.

DRILLING MACHINE – Machine tool used for the production of holes with a drill, fall under two classes—bench drilling machine and the pillar drilling or upright drilling machine.

DROP STAMP – Machine for producing drop stampings and forgings, may be free falling type or controlled falling type.

DRY CELL (dry battery) – A battery that uses no liquid electrolyte.

DUPLEX MILLER – Manufacturing type milling machine sometimes called a duplex head miller, has two head stocks supporting two opposed spindles. This enables two face mills or two horizontal cutters, to work simultaneously on opposite sides of the workpiece.

DYNAMOMETER – A device for absorbing the power output of an engine and measuring torque or horse power, so that it can be computed into brake horsepower.

ELECTRIC DRILL – Hand operated, portable drilling machine driven by electric power.

ENGINE – A machine that converts heat energy into mechanical energy. The assembly that burns fuel to produce power is sometimes called POWER PLANT.

FIXED DISPLACEMENT PUMP – A type of pump in which the volume of fluid handled per cycle cannot be varied.

FINE BORING MACHINE – Machine used for boring to very close limits, the holes in pistons, the bores in cylinder blocks, the bearings in motor car and aero engine connecting rods, valves as fitted to steam and other piping, may have a horizontal or vertical spindle.

FLY BALL GOVERNOR – Conventional type of centrifugal governor, commonly called a mechanical governor.

FORGING MACHINE – Machine designed to shape a metal article while the material is in the hot plastic state by applying the necessary force or pressure.

GANG DRILLING MACHINE – A drill press in which two or more drilling spindles are mounted on a single table. This permits several drilling operations to be made by moving the workpiece from spindle to spindle in a planned sequence.

GAUGE SNUBBER – A device installed in the fuel line connected to the pressure gauge, used to dampen pressure surges and thus provide a steady reading. This helps protect the gauge.

GEAR TYPE PUMP – A pump which uses the space between the adjacent teeth of gears for moving the liquid.

GENERATOR – An electromagnetic device used to generate electricity.

GOVERNOR – A device for controlling the speed of a prime mover, that produces power.

GRINDING MACHINE – Machine designed for one or other or a combination of the following functions. Grinding excess material off parts previously machined by other methods; forming parts from blanks; parting off lengths of material from rods, bars, billets and so on.

HEAT EXCHANGER – A device used to cool a fluid by transferring heat.

HOIST – Appliance or machine for lifting heavy material, also called lifting tackle.

HYDRAULIC GOVERNOR – A governor using fluid to operate the fuel control.

INTERCOOLER – Heat exchanger for cooling the air between the stages of compression.

INTERNAL COMBUSTION ENGINE – An engine that burns fuel within the cylinder itself as a means of developing power.

INTERNAL GRINDER – Machine used to finish internal surfaces such as engine cylinders, ball bearing races, and similar work that requires excellent accuracy and high finish.

JIG BORING MACHINE – An extremely accurate machine tool having a precession spindle and a table to hold the work. The table and spindle are movable with built in measuring devices. Used for spacing and producing precision holes in tools, dies, and small lot production where extreme accuracy is required.

KNUCKLE PRESS – Press designed to exert a great pressure. It is a strong, heavily built and a rather slow-moving machine, and the standard range varies from 60-100 tons per sq.inch.

LEVER – A simple machine for obtaining mechanical advantage. The lever consists of a rigid arm or bar pivoted or bearing on a point called the fulcrum and has a weight arm and a power arm.

LINCOLN MILLER – One of the manufacturing type milling machines, all of which are characterized by the fixed table height with vertical adjustment of the spindle, in contrast with the column and knee type in which the work table is vertically adjustable.

LINISHING MACHINE – Automatic mirror polishing machine capable of polishing one or more parts at a time on the same polishing band or disc.

LIQUID COOLED ENGINE – An engine that is cooled by the circulation of liquid coolant around the cylinders.

LUMSDEN GRINDER – Oscillating type of grinding machine. The tool to be ground is secured in a chuck having a range of adjustments such that any desired angle can be ground.

MACHINE – A mechanical device, mechanism, or apparatus which transmits power and motion to perform work.

MACHINE TOOL – Type of power driven machine for cutting metal as employed in machine shops, the term being usually restricted to those which use the tool that produces waste material in the form of chips or cuttings. Typical machine tools are lathes, planers, shapers, slotters, borers, drills, broaches, millers and saws.

MANDREL PRESS – Press for the purpose of fitting mandrels to the bore of the work.

MICHIGAN GEAR SHAPER – Machine incorporating a number of shaped cutters, used to produce all the teeth on spur or helical gears simultaneously.

MILLING MACHINE – Machine tool designed to remove metal from the surface so as to size and shape it by means of a revolving multiple toothed cutter called milling cutter.

MOTOR – A rotating electromagnetic device used to convert electrical energy into mechanical energy.

MOTOR VEHICLE – Any type of self propelled vehicle mounted on wheels or tracks.

MOULDING MACHINE – Mechanical device for making moulds at a much faster rate.

MULTIPLE SPINDLE DRILLING MACHINE – Machine used for drilling a large number of holes in one operation, built purely for repetition work.

MULTIPLE SPINDLE LATHE – Machine, generally of the automatic type, with up to six spindles mounted in a drum, this indexes in sequence, so that a different operation is performed at each station, and one revolution of the drum completes a piece of work.

NIBBLER – Machine designed to cut metals in sheet form to any required shape. The machine comprises a small punch moving up and down at high speed, and so taking a considerable number of small nibbles or cuts.

NUT TAPPING MACHINE – Device for tapping, that is providing the screw threads on nuts.

OIL COOLER – A heat exchanger for lowering the temperature of oil.

OIL PUMP – Mechanical device to pump oil (under pressure) into the various oil galleries.

OPTICAL COMPARATOR – A machine that enlarges the profile view of a part and compares it to a standard profile.

OSCILLOSCOPE – A device for recording wave forms on a fluorescent screen, proportional to the input voltage.

OVER RUNNING CLUTCH – A clutch mechanism that transmits power in one direction only.

OVER SPEED GOVERNOR – A governor that shuts off the fuel or stops the engine, only when excessive speed is reached.

PLAIN MILLER – Milling machine in which milling can be carried out in a direction parallel with the axis of the arbor or at right angles to it , but no swivelling of the worktable is possible.

PLANING MACHINE – Machine for removing material from plane surfaces by means of cutting tools fixed in the machine. It comprises a bed, a travelling worktable on which the parts to be planed are carried, standards, a cross slide, tool box and gearing. Also called PLANER.

PLANOMILLER – Milling machine, sometimes called a slab miller, similar in shape and appearance to a planing machine, but having cutter hands instead of clapper type tool boxes.

PLUNGER PUMP – A pump which operates and displaces fluid by means of a plunger.

PNEUMATIC TOOL – Design of hand tool operated by air pressure *e.g.*, pneumatic drill, pneumatic hammer or riveter.

POWER HAMMER – Machine for shaping metal while hot, grouped into three classes, according to the power supplied: compressed air, steam or hydraulic pressure.

POWER PLANT – The engine or power producing mechanism.

PRESS – Machine designed to manipulate sheet and strip metal so as to produce a given shape or form.

PROFILING MACHINE – Form of milling machine designed to produce delicate parts whose form or contour is irregular, particularly for mass production by the use of jigs and so on, of small machine and other parts, or where the production of interchangeable parts is desired.

PRONY BRAKE – A friction brake used for engine testing.

PUMP – Mechanical device in which a fluid, gas or liquid, is lifted or forced against an external pressure, main types are reciprocating or piston pumps, centrifugal pumps and rotary displacement pumps.

PUNCHING MACHINE – Machine designed to cut metal blanks from the sheet. It may be of the crank or eccentric variety; the frame of the C or double sided type. Also called PUNCH PRESS.

RADIAL DRILLING MACHINE – Machine with the drill head working on an arm which has a rotary movement in the horizontal plane. In this the drill spindle could be moved to the positions required instead of moving the work for each hole to be drilled.

RADIATOR – A heat exchanger in which cooling water gives up heat without coming into direct contact with air.

RAM (pump) – Hydraulic ram, also termed a water ram, by which water is raised by its own momentum from a lower to a higher level.

RECTIFIER – A device, made of a diode or series of diodes, used in a welding machine, for converting alternating current to direct current.

RELAY – An electromagnetic switch which utilizes variation in the strength of an electric circuit to effect the operation of another circuit.

RELIEF VALVE – An automatic valve which is held shut by a spring of correct strength. Excessive pressure opens the valve and releases some of the gas or liquid.

RELIEVING LATHE – Machine for backing off the teeth of milling cutters to facilitate machining of metals.

REVERBERATORY FURNACE – Furnace in which ore or metal is melted by contact with the flame, but is not in contact with the fuel.

RHEOSTAT – A device to regulate current flow by varying the resistance in the circuit.

RIGBY HAMMER – Steam hammer used for a wide range of forging work.

RIVETING MACHINE – Power operated machine used to form rivet heads. May be classified according to the method of operation hydraulic, pneumatic, or steam and the manner in which the rivet head is formed.

ROLLING MILL – Mill in which ingots, bars, blooms, billets etc., of steel or other metals, are rolled to reduce or alter their sections.

ROLL TURNERS LATHE – Special design of lathe adapted for machining rolls employed for printing machines, calenders or rolling mills.

ROOTS BLOWER – An air pump or blower similar in principle to a gear type pump.

ROTARY BLOWER – Pump or compressor used to supply air or gas to a furnace, forge, internal combustion engine or similar application. This may be a vane, geared rotor, centrifugal or turbine type depending on the particular application and the volume of air or gas or the pressure required.

ROTARY PUMP – Pump used for liquid or gas, in which a rotating impeller is used, may be vane type, gear type centrifugal and turbine type.

SALT BATH FURNACE – Liquid bath furnace used for case hardening and heat treating of metals and alloys.

SAWING MACHINE – Machine tool for cutting metal by the use of a blade band, or disc provided with cutting teeth.

SCAVENGING PUMP – A piston type of pump delivering scavenging air to an engine.

SCRAP CUTTER – A type of double action press having a side ram whose purpose is to cut the scrap metal into short lengths in order to facilitate handling.

SCREW MACHINE – A power driven automatically controlled machine tool designed to produce finished parts from bar stock. Cams control the movement of the cutting tools, attachments and stock feeding devices to complete workpieces automatically at high production rates.

SENSITIVE DRILLING MACHINE – Type of machine in which drills are used with a sensitive hand pressure action.

SHAPING MACHINE – Type of machine used for the production of flat or curved surfaces. In this the ram which carries the tool reciprocates while the worktable which carries the workpiece moves perpendicular to the ram movement. Also called SHAPER.

SHEARING MACHINE – Machine designed for cutting sheet metal. It comprises a fixed blade and a shearing tool or blade.

SILENCER – A device for reducing the noise of intake or exhaust in an engine.

SLOTTING MACHINE – Machine whose mechanical principles are much the same as those of the shaping machine, but with the difference that the ram on which the slotting tool is mounted moves vertically, as in a vertical shaping machine. Also called SLOTTER.

SOLENOID – An electrically operated magnetic device used to do work.

SPOT WELDING MACHINE – Fixture in which sheets of metal to be welded together are clamped between two electrodes through which an electric current is passed.

SUPERCHARGER – An air pump driven by an engine, which fills the cylinders with air at a higher pressure than atmospheric pressure.

SURFACE GRINDER – It is a general purpose machine tool used to finish machine a plane surface. There are two types—horizontal and vertical.

TAPE CONTROLLED MACHINES – Machines which have partial automation in which the movements of a machine tool and cutters are automatically controlled by a magnetic tape or cards which have been punched to indicate data previously planned for the sequence of each operation to be performed.

TAPPING MACHINE – Machine designed to cut threads in holes in nuts and other parts.

TESTING MACHINE – Machine designed to apply accurately measured loads to a test piece and, in many cases, to measure the resulting extension, compression or deflection, in order to ascertain the physical properties of materials.

THYRISTOR – A type of semiconductor device that acts as a switch. It turns on when a certain voltage is applied to the gate, and it turns off when the current flowing between the other two terminals stops or reverses.

TOOL GRINDER – Special machine for the grinding of turning, planing and other tools.

THREAD MILLER – A machine that cuts threads by using a rotating milling cutter.

TROCHOID PUMP – An oil pump consisting primarily of a pair of rotors of special shape.

TURBO CHARGER – An exhaust gas driven turbine directly coupled to a compressor wheel.

TURRET LATHE – Machine generally larger than a capstan lathe, but using similar tools. The main difference between them is that the turret saddle has longitudinal movement on the guides of the bed.

UNIVERSAL GRINDER – Type of grinding machine having a wide range of application and specially suited to work that has to be ground at an unusual angle. Provided with a swivel table and headstock and a wheel head that can be rotated on its base.

UNIVERSAL MILLING MACHINE – Design of milling machine of the column and knee type, roughly similar in general design to plain milling machine (column and knee type), but usually of less robust construction and having additional features that widen its field of usefulness *e.g.*, a worktable that swivels in a horizontal plane so that it is not restricted to two right angled motions.

VERTICAL MILLING MACHINE – Milling machine in which the cutter spindle rotates about a vertical axis (as on most drilling machines), but which in certain other respects has much in common with a plain miller.

VERTICAL TURRET LATHE – A lathe with a vertical spindle and ways that carries the work in a chuck and has cutting tools mounted in a five position turret and side heads.

VIBRATION DAMPER – A specially designed device mounted to the front of an engine crank-shaft to reduce torsional vibration.

WANKEL ENGINE – A rotary type engine in which a three lobe rotor turns eccentrically in an oval chamber and thus performs cycle of operations.

WATER BRAKE – A device for engine testing in which the power is dissipated by churning water.

WELDING MACHINE – Equipment used to perform the welding (joining) operation, for example, spot welding machine, arc welding machine, seam welding machine etc.

CHAPTER 6

MACHINE COMPONENTS AND ATTACHMENTS

ANGLE PLATE – Right angled metal plate used to secure parts during machining or when taking measurements.

ANVIL – Heavy block on which to hammer and shape metals.

APRON – The function of a lathe apron is to carry the mechanism for sliding and surfacing motions and screw cutting.

BACK GEARS – Gears applied to machine tools to increase the number of speed changes obtainable with a cone or step pulley belt drive.

BASTERED THREAD – A screw thread which does not confirm to any recognized standard dimensions.

BED – One of the principal parts of a machine tool having accurately machined ways or bearing surfaces for supporting and aligning other movable parts of the machine.

BELLCHUCK – Hollow cylindrical chuck bolted to the main chuck for the purpose of giving additional support to work of awkward shape.

BELT SHIFTER – A flat hardwood strip of suitable length having shifter fingers attached at one end and used to shift a belt from one pulley to another or to replace a belt which has run off a pulley on an overhead drive shaft.

BOLSTER – Support for dies and tools in forging presses and drop stamps.

BOX ANGLE PLATE – An angle plate made of cast iron, usually with slots cast in it and accurately machined on the outside.

BOX JIG – A jig made in the form of a box into which the job to be drilled is inserted.

BULL WHEEL – The large gear wheel of a planer which meshes with the rack under the table and drives it. The large crank gear of a shaper is often called a bull wheel.

CAP SCREW – A finished screw, used for fastening two pieces together by passing the screw through a clearance hole in one part and screwing it into a tapped hole in the other. Heads may be hexagon, round, flat, fillister or socket type.

CARRIAGE – A principal part of lathe. The carriage carries the cutting tool and moves along the ways of the lathe. It contains the saddle, compound slide, and apron.

CARRIER – Tool for driving work which is held between centers, as in a lathe.

CATCH PLATE – Plate screwed to the nose of the lathe spindle for the purpose of driving work held between centers, through the medium of a carrier or driving dog.

CENTRE SQUARE – Device to enable the rapid location of the centre of the flat end of a cylindrical workpiece.

CHANGE GEARS – An assortment of gears which are supplied with a machine for changing speed ratio between driver and driven parts of the machine. Change gears on a lathe make it possible to cut threads of different pitches and obtain different feeds per revolution. On milling machines they are used to obtain different leads when milling spirals and helices.

CHUCK – Appliance for gripping tools, such as drills, or for holding work in a lathe.

CLAPPER BOX – A part of the shaper tool head that holds the tool post. A clapper block is hinged into the clapper box to permit the cutting tool to swing upward on the return stroke.

COLLET – Means of gripping a bar to give quicker chucking, particularly in capstan work for rapid and accurate setting.

CLAMP – Device for holding work during marking out, measuring, machining, fitting or grinding.

COMPOUND SLIDE – A principal part of a lathe, frequently called a COMPOUND REST, consisting of an upper and lower part dovetailed together. The lower part or base is graduated in degrees and can be swivelled to any angle for turning short tapers and angles. The upper slide carries the tool post and tool holder.

CONE PULLEY – A stepped pulley having two or more diameters and made in one piece.

DEAD CENTER – The center fitted to the tailstock of a lathe and so named because it does not rotate. Also the exact or precise center of an object.

DEMAGNETIZER – Device for removing permanent magnetism from steel tools and parts.

DIE HEAD – Device to enable small or medium sized work to be screwed in capstan or turret lathes. The die head consists of a cylindrical body with a shank to enable it to be clamped in the turret.

DIE SCREW – Tool used when threading cylindrical pieces beyond the capacity of a die plate.

DISTANCE BLOCK – Accurately ground steel block used to hold a component at a set distance from another surface, or to operate a trip mechanism tool when the tool carrier or worktable reaches the desired position.

DIVIDING HEAD – Indexing or dividing head is an attachment used on the milling machine table, for accurately dividing the circumferences of components for grooving or fluting, gear cutting, cutting of splines, squares or hexagons.

DOG – A tool or accessory which can be clamped on a workpiece to drive it while held between centers, as a lathe dog. Also, the name given to a projecting part on a machine tool which strikes and moves another part, such as the reversing dogs on a grinding machine or planer.

DRIFT PIN – A round tapered steel pin used to align rivet holes so that the rivet will pass through the holes easily.

DRILL ADAPTER – An attachment used to fit square shank drills and morse taper sockets.

DRILL CHUCK – A self centering tool holding device, for driving drills and other such cutting tools in a drilling machine.

DRILL DRIFT – A wedge shaped piece of steel that is used to remove tapered shank tools, such as drills from spindles, sockets and sleeves.

DRILL HOLDER – An attachment used for holding drills in the turret of a capstan lathe.

DRILLING JIG – Device, usually of hardened steel plate, for holding a piece of work in a definite position, provided with guide bushes through which drills can pass so as to locate holes in exact positions in the work.

DRILL SLEEVE – A steel part tapered inside and outside which fits onto the tapered shank of a cutting tool, such as a drill or reamer, to adapt it to the size of the hole in the drill press spindle.

DRILL VISE – A work holding device in which the work is set and held while drilling and other operations are performed.

DRILL SOCKET – An adapter for holding taper shank drilling tools which have a taper shank larger than the taper hole in the drilling machine. One end is tapered outside to fit the drilling machine spindle hole while the opposite end has a larger taper hole to receive the cutting tool shank.

EXPANDING MANDREL – It is used to hold work on an internal diameter. It is similar in principle to a spring collet. Three jaws are arranged to expand outward and clamp the work, as a tapered pin is forced inward.

FACE PLATE – Flat, slotted plate screwed on the nose of the lathe spindle, or bolted to a flange, to enable work to be machined which cannot be held in the chuck.

FEATHER – A sliding key sometimes called a spline, used to prevent a pulley, gear, or other part from turning on a shaft, but allows it to move lengthwise. The feather is usually fastened to the sliding piece.

FEED STOP – Method of tripping the power feed of a machine tool at a predetermined point.

FIXTURE – Device by which a component to be machined is correctly located and held during a machining operation.

FLASK – Complete moulding box that contains the sand that form the mould.

FOLLOWER REST – A support for long, rather small diameter work to be turned in the lathe. The rest is attached to the carriage and set close to the cutting tool. It travels along with the cutting tool to prevent the work from springing away from the cutting tool.

FOOT STOCK – Part of a dividing head attachment for the milling machine. It contains a center for supporting the workpiece, the same as the tailstock does for a lathe.

FOUR JAW CHUCK – A chuck that provides a more powerful grip on the work by means of four jaws which may be adjusted independently, so that irregular shapes may be gripped, and the fact that the jaws can be reversed enables the work to be gripped inside as well as from the outside.

GIB – An angular or wedge like strip of metal placed between two machine parts, usually sliding bearings to ensure a proper fit and provide adjustment for wear.

GOGGLES – These are worn when tool grinding, to protect the eyes from the flying particles of grit.

HALF NUT – A mechanism attached to the apron of a lathe and operated by a lever. The lever opens and closes a nut that has been split lengthwise so that the two halves of the nut can be closed upon the lead screw when threads are being cut. Also called SPLIT NUT.

HAND VICE – Small vice used for holding light articles such as pins, screws etc., especially during heat treatment.

HAND WHEEL – Any of the wheels found on machine tools for moving or positioning parts of a machine to hand feed, as the tailstock handwheel on a lathe.

HARDIE HOLE – Square hole in the face of an anvil which forms the location for the hardie and other tools used by the smith, such as small swages, fullers and cutters.

HEAD STOCK – An assembly containing the headstock spindle and the mechanism for driving it which is fastened permanently to the left end of the lathe bed.

HOLDING BLOCK – Fixture designed to hold small or thin gauge parts which cannot readily be held in a vice.

IDLER GEARS – A gear or gears placed between driver and driven gears when the distance between the driver and driven gears would require too exceptionally large gears. These gears have no effect on the speed or ratio between the driver and driven gears.

INDEPENDENT CHUCK – A three jaw chuck, in which each of , the jaws may be moved independently of the other two.

INDEX CRANK – The crank handle of an index or dividing head used for turning the spindle through a full turn or fractional part of a turn. A pin on the crank fits into the holes of the index plate.

INDEX PLATE – A thin circular plate having various circles of holes. It is attached to the index head and used for obtaining fractional parts of a turn of the index crank. A set of three or more index plates may be furnished for an index head.

INTERMEDIATE GEAR – An idler or loose gear on a stud for transmitting power between a driver and a driven gear.

JACOB'S CHUCK – Gear operated drill chuck of three jaw type for use on drilling or milling machines and lathes.

JACOB'S TAPERS – Standard system of tapers which are used for drill chucks and spindles.

JIG – A device that locates and holds a workpiece and guides the cutting tool.

JIG BUSH – Hardened steel cylindrical guide, made a press fit into the body of the jig for a drill or reamer to pass through.

JIG PLATE – Flat steel plate provided with a number of bushed holes and used to locate the drill when drilling a component.

JOURNAL BOX – A housing or support for a bearing which contains the journal of the shaft.

KEYS – Metal pieces of various designs which fit into a slot in a shaft and project above the shaft to fit into a mating slot in the center hole of a gear or pulley to provide a positive drive between the shaft and the gear or pulley.

KNEE – A principal part of a column and knee type milling machine which slides vertically on the column and carries the saddle and table.

LADLE – Receptacle used for transporting molten metal.

LATHE CARRIER– It is clamped on a bar and engaged with a projecting dog on the catch plate fitted to the lathe head stock. This enables the bar to revolve between the centres.

LATHE CENTERS – Solid steel pieces which have a tapered shank and 60° pointed end. A live center fits into the headstock spindle and a dead center fits into the tailstock spindle to support and provide bearing points for the work turned between centers.

LAYOUT PLATE – A smooth finished flat steel plate on which work and tools may be placed and used while laying out a job.

LEAD SCREW – Master screw used for cutting a screw thread in a lathe.

LIVE CENTER – A lathe center that fits into the headstock spindle and so named because it revolves.

LEVER – A simple machine for obtaining mechanical advantage. The lever consists of a rigid arm or bar pivoted or bearing on a point called the fulcrum and has a weight arm and a power arm.

LOCK NUT – A type of nut that is prevented from loosening under vibration. The locking action is accomplished by squeezing, gripping or jamming against the bolt threads.

LOOSE PULLEY – A pulley which turns freely on a shaft so that a belt can be shifted from the driving pulley to the loose pulley in order to stop a machine driven by an overhead belt drive.

MACHINE VICE – A work holding device used on machine tools to seat and hold work securely in position while machining is done.

MAGNETIC BLOCK – Laminated block for use in conjunction with a magenetic chuck.

MAGNETIC CHUCK – Magnetic device for holding work during machining and grinding.

MANDREL – A turned, hardened and accurately ground cylindrical bar or shaft or spindle, incorporating a slight taper, used for holding bored parts in a lathe while the outside surface is being turned.

MANDREL PRESS – Press for the purpose of fitting mandrels to the bore of the work.

MANIPULATOR – Fixture in which work is held for welding.

MILLING HEAD -- An attachment or device for the milling machine to add to the range of operations for increasing production, versatility, and accuracy of the machine. A vertical attachment makes it possible to do vertical milling on a horizontal machine.

MITER GEARS – Bevel gears which have a center or pitch angle of 45°.

MORSE TAPER – Standard system of tapers widely used on lathes and drilling machines.

MULTIPLE THREAD SCREW – A screw with two or more threads cut around the periphery of the workpiece to provide an increased lead with a specified pitch.

NUT – A metal fastener of square, hexagon or other shape having an internal thread which screws onto a bolt, stud, or arbor.

NUT BOX MECHANISM – In a lathe it is used to engage and disengage the lead screw for thread cutting. This comprises a pair of half-nuts capable of sliding in vertical slides in or out of mesh with the lead screw.

OVERARMS – Adjustable supports for the end of a milling cutter arbor farthest from a milling machine spindle.

PARALLEL JAW VICE – Type of vice in which the gripping faces between the sliding and fixed jaws are always parallel, irrespective of the gap between them.

PILOT – A guide at the end of a counterbore which fits freely into the drilled hole to hold and align the body of the counterbore while cutting takes place.

PINION – The smaller of a pair of gears regardless of the size or type.

PIN VICE – Small vice with self centering jaws useful for gripping small rods, wires, drills and files.

PIPE THREAD – A 60° thread form having flattened crests and roots which are cut on a taper having 3/4 inch taper per foot. Pipe thread is used on piping and tubing.

PIPE VICE – Clamping device designed to hold work of tubular or circular section.

POWER ASSISTED CHUCKS – These are used where great holding power is required. These are operated with auxiliary power such as compressed air or oil pressure. Where simplicity of chucking is the first consideration, magnetic chucks are used.

PULLEY – A wheel having a plain or V groove rim over which a belt runs for the transmission of power from one shaft to another.

QUICK CHANGE GEARBOX – A cluster of gears on a machine tool arranged in such a way that a sliding gear can be moved by a lever to mesh with anyone of the gears in the cluster to obtain desired ratio of speeds, feeds or thread pitches.

QUICK RETURN MECHANISM – A mechanism on various machine tools to give a table, ram or other movable part a rapid movement during the return or non-cutting stroke.

QUILL – A hollow shaft that revolves on a solid shaft, carrying pulleys, gears or clutches. When the clutch is closed, the quill and shaft revolve together.

RACK – A straight metal strip having teeth that mesh with those of a gear to convert rotary motion into reciprocating motion or just the opposite.

RATCHET – A gear with triangular shaped teeth to be engaged by a pawl which gives it intermittent motion or locks it against backward movement.

RELIEVING ATTACHMENT – A device or mechanism fastened to the cross feed of a lathe to provide a reciprocating motion to the cutting tool, to obtain a clearance in the back of the cutting edges of the teeth when making taps, milling cutters or hobs.

REVERSE LEVER – A lever on the machine tool for changing the direction of some movable part such as the lead screw on a lathe.

REVOLVING LATHE CENTRE – It is fitted in the tailstock instead of the fixed type of loose head centre, in some cases, so that the high speeds of work revolutions become possible.

RIGHT HAND THREAD – A screw thread which advances into the mating part when turned clockwise or to the right.

RIVET – A one piece fastener consisting of a head and a body and used for fastening two or more pieces together by passing the body through a hole in each piece and then forming a second head on the body end. It cannot be removed except by taking off the head.

ROLLER STEADY – Type of steady fitted with roller points to support long or slender work in a lathe.

ROTARY TABLE – An attachment for the milling machine consisting of a round work holding table provided with T slots and operated by a handwheel attached to a worm or worm gear for rotating the table. A scale graduated in degrees permits the work to be moved any number of degrees for spacing holes or milling radii. Also called CIRCULAR MILLING ATTACHMENT.

SCREW – A helix formed or cut on a cylindrical surface which may advance along the axis to the right or left. The helix may be single or multiple.

SCREW JACK – Device used for lifting a heavy load by means of a screw.

SCREW THREAD – A ridge of uniform section or shape in the form of a helix on the external or internal surface of a cylinder, or in the form of a conical spiral on the external or internal surface of a cone.

SELF CENTERING CHUCK – Design of chuck for holding hexagonal or cylindrical work, operating in such a way that the work is automatically centered.

SET SCREW – Usually a hardened steel screw having either no head or a square head and with various designs of points or ends to lock or tighten adjustable machine parts in position on a shaft.

SHOULDER SCREW – A screw having two or more diameters or shoulders and commonly used for supporting levers and other machine parts that have to operate freely. The screw body is slightly longer than the thickness of the piece pivoted on the screw so as to allow the piece to move freely when the screw is set up tightly against the bottom of the shoulder.

SLEEVE – A round piece of metal having a straight or tapered hole which fits over or into another piece to adapt parts to fit, as a taper sleeve for a lathe center.

SLIDE REST – A tool holding slide used on machine tools. It is usually adjustable for holding and guiding cutting tools. The compound slide or rest used on an engine lathe is an example. The compound slide permits the tool to be fed at any angle.

SLIDING TABLE (**compound**) – Device to enable holes to be drilled or bored very accurately with respect to each other.

SNAP FLASK – Strongly constructed wooden moulding box, split diagonally with hinges on one corner and locking latches on the opposite corner, so that the box may be removed from the mould prior to casting.

SOCKET HEAD – Screw threads having a hexagonal or other form of recessed socket in the head so that the screw can be turned with a wrench or key, as a hexagonal key.

SPLINES – Multiple keys cut in either a shaft or a hole of parts that are made to slide or move lengthwise on a shaft, as a sliding gear transmission.

SPLIT NUT – A nut which has been cut length wise so as to open for quick adjustment.

SPRING COLLETS – A type of draw in collet made of hardened steel and having three slots or saw cuts which permit the collet to be closed tightly against a tapered sleeve by the draw bar.

SQUARE THREAD – A form of screw thread in which the cross-section of the thread forms a square, making the width of the thread equal to the width of the space between the threads.

STAKE – Small anvil used for sheet metal work.

STEADY REST – A support attached to the ways of a lathe for turning long workpieces. Used to prevent slender work from springing away from the cutting tools or to permit machining operations to be performed on the end of the workpiece. Some times called a center rest or STEADY.

STEP BLOCK – A block of steel or cast iron having a series of steps and used for supporting the ends of machine clamps when clamping work to the table.

STEPPED CONE PULLEY – A cone pulley having several diameters or steps for obtaining different speeds on machine tools.

STOPS – Devices attached to the movable table or ram or parts of a machine tool, such as a milling machine table, to limit the amount of travel.

STUD – A plain cylindrical piece having a threaded portion of suitable length at each end or a continuous thread over the entire length. One end is screwed into a machine or workpiece after which a second part is placed over the stud and held in place by a nut.

SWIVEL SLIDE – Device to enable the toolpost slide of a lathe to be set at an angle for taper turning or cutting V threads.

SWIVEL VICE – A bench or machine vice for holding workpiece which may be rotated on its base to the desired angle or position.

TABLE (marking off) – Table used for marking off, or setting out a job to be machined.

TAILSTOCK – The principal part of the engine lathe, used for supporting one end of a workpiece by means of a center point held in the spindle. May be moved along the ways and clamped in different positions and offset from the true axis of the lathe for turning tapers.

TAPER PINS – Steel pins used for locating and holding machine parts in position on a shaft.

TAPER TURNING ATTACHMENT – Device fitted to a lathe, to cause movement of cutting tool at an angle to the axis of the lathe so as to produce the required taper on the workpiece.

TAYLOR CHUCK – Special chuck designed so that the jaws are supported at right angles to the direction of the pressure.

T-BOLT – A threaded bolt having a square or rectangular end which fits into the T slot of a machine table for clamping workpieces.

THREAD CHASING DIAL – A device consisting of a short shaft with a dial at the top and a worm gear at the bottom to mesh with the lead screw on an engine lathe which is attached to the lathe carriage, so that the operator can engage the split nut lever at the correct position when cutting screw threads.

THREE JAW CHUCK – Chuck for gripping general work using three jaws, usually concentrically, simultaneously, a feature which makes the chuck particularly suitable for holding either round or hexagon stock.

THUMB SCREW – A type of screw having a winged or knurled head for turning by hand when a quick and light clamping effect is desired.

TOOL HEAD – The part of machine tool that carries the cutting tool and is adjustable for moving and setting the cutting tool into the workpiece being machined.

TOOL HOLDER – Slotted shank in which the tool bit can be held.

TOOL POST – A clamping device for holding tool holder in position on a machine tool.

TOP SLIDE – Means of traversing the tool on the tool post of a lathe.

T-SLOT – A recessed or undercut slot made with a special T shaped cutter in the tables of machine tools to receive the square head of a T bolt for clamping workpieces.

TUMBLER GEARS – Two gears acting as idler gears between the spindle gear and the fixed stud gear on a lathe gear train. Used to reverse the direction of rotation of the lead screw for thread cutting or the feed rod for turning.

TURRET – Rotatable device for holding several cutting tools set in position to operate in turn.

TWO JAW CHUCK – Chuck furnished with two jaws, diametrically opposite.

U-BOLT – An externally threaded fastener bent in the shape of the letter U and with both ends threaded.

U-CLAMP – A flat or square piece of metal bent or formed into the shape of the letter U and used for clamping workpieces on a machine table.

UNIVERSAL CHUCK – Design of chuck in which the jaws are moved simultaneously by gear wheels, by a scroll or by a combination of both. Also called concentric chuck or self centering chuck.

UNIVERSAL INDEX CENTERS – A dividing head and foot stock with a set of change gears to gear the spindle of the index head to the index crank for differential indexing, and to gear the index head to the lead-screw of the table of a milling machine for helical milling.

UNIVERSAL VICE – A work holding device usually used on the milling machine which has either two or three swivel settings so that workpieces may be set at any desired angle including compound angles.

V-BLOCK – Block of metal having a V-shaped groove cut in its top surface in which can rest a circular shaft.

V-SUPPORT – Steady in the form of a V, to support the work near the cutting tool.

VICE – A mechanical device of many designs and sizes in which workpieces are clamped for hand and machine operations.

VICE CLAMP – Piece of soft material so shaped as to enable it to fit over each of the hardened jaws of a vice in order that workpieces may be gripped without damage.

V-WAYS – The top part of the bed of a lathe, planer or other machine bed which acts as a bearing surface for aligning and guiding the moving parts, such as the carriage of a lathe.

WAYS – The flat or V shaped bearing surfaces on machine tools over which other movable parts slide.

WOODRUFF KEY – A flat semicircular piece of metal used as a key in a special circular key seat slot cut in a shaft to drive a gear, pulley or other part.

CHAPTER 7

TOOLS AND CUTTERS

ABOUTSLEDGE – The large hammer used by a blacksmith's mate, turn-about with the smaller hammer of the blacksmith.

ANGLE CUTTER – Type of milling cutter used for single or double angles. The term angle cutter covers three types of milling cutters, the single or half-angle, the double angle and the equal or combined angle.

BACK STICK – Piece of wood used when spinning sheet metal by hand in a lathe.

BASTERED FILE – File of approximately middle grade in regard to cut or tooth pitch.

BELL CENTER PUNCH – Device used for rapidly locating and marking the centre of the flat end of a cylindrical workpiece, preparatory to heavier centre punching, centre drilling, turning in a lathe etc.

BLANKING AND CUPPING TOOL – Tool used to cut a blank and form a cup from sheet or strip metal at one stroke of the press.

BLOW PIPE – Gas welding torch in which oxygen and acetylene are mixed and ejected from a nozzle.

BOILER TAP – Hand tap specially designed for tapping holes for use with boiler stays.

BOLSTER – It is a block of mild steel with a hollow in it to accommodate the rivet head.

BORING BAR – Bar carrying a cutter or cutters to enable holes to be bored which are larger in diameter than can be conveniently drilled by means of twist drills.

BORING TOOL – Single or double ended tool for machining a drilled or cored hole.

BOX SPANNER – Spanner in the form of a hollow tube, shaped at the end (or ends) to fit a nut.

BOX WRENCH – A type of closed end wrench made in many styles for specific sizes and shapes of bolt heads or nuts.

BREAST DRILL – Has an adjustable breast plate, by means of which a much greater pressure may be applied to the drill. This hand-drill has double ratio drive.

BROACH – A long tool with number of cutting teeth which is pushed or pulled through a hole or across a surface to form the desired shape and size.

CAULKING TOOL – Type of tool used to close the riveted joints on pressure vessels, for instance, a boiler, or compressed air reservoir.

CEMENTED CARBIDE TOOL – Lathe or other cutting tool fitted with a tip made from either tungsten, tantalum or titanium carbide, held in a matrix of cobalt.

CENTER DRILL – A short, stubby, straight shank, two fluted twist drill that is used when center holes are drilled and reamed in the ends of a shaft in separate operations.

CENTER POP – A dot punch or sharply pointed center punch.

CENTER PUNCH – Hand tool employed for marking out points or centers *e.g.*, when marking the center of a shaft preparatory to drilling or center drilling.

CENTER REAMER – A countersink having a 60° included angle for sizing and smoothing center holes in workpieces to be turned or ground between centers.

CERAMIC TOOLS – A newer cutting tool material made of aluminium oxide or silicon carbide and held together by binders or additives of other materials.

CHAIN CUTTER – Cutter of chain type used in a chain mortise machine for the mortising of wood.

CHASER – Tool for forming or finishing screw threads in a lathe.

CHIP BREAKER – A groove ground into the top of the toolbit of a lathe, shaper or planer to keep the chips short.

CHISELS – These are hand working tools, made from carbon tool steel, usually of octagonal section. The end of the chisel is shaped to the required form and finished by grinding.

COMPARATOR – Instrument for comparing the dimensions of a component with a standard of length.

COMPASS – Instrument used for describing arcs and circles.

COTTER MILL – Also called slotting end mill, cottering or key seating cutter.

COUNTERBORE – When a drilled hole is to be enlarged concentrically for a portion of its length, a counterbore is used. It consists of a series of blades arranged to cut on the ends only, and a smooth pilot slightly smaller than the drilled hole.

CROSS CUT FILE – File which is parallel throughout its length and in section tapers more sharply than a knife file, used extensively for sharpening the teeth of cross cut saws.

CROSS CUT OR CAPE CHISEL – It is forged so that the cutting edge is slightly wider than the body, this is to ensure that the chisel does not bind in the cut when it is used for deep grooves.

CROSS PEEN HAMMER – A hammer with a wedge shaped peening end at right angles to the direction of the handle.

CURLING TOOL – Tool designed and made to curl the edge of a metal article.

DEAD SMOOTH FILE – A file having teeth of very fine cut in other words, with its teeth very closely spaced or pitched.

DIAMOND POINT CHISEL – A type of cold chisel ground with a diamond point.

DIAMOND WHEEL – An abrasive wheel composed of a bonding agent, into which crushed and screened diamond particles of appropriate size are introduced to form the abrasive cutting face.

DIE (drawing) – Tool of angular form through which a shell is pushed, tubes, sections, bar and wire are pulled so as to give the required size and cross-sectional shape.

DIE NUT – Tool, shaped like a square or hexagonal nut, and provided with cutting edges, employed for cutting or sizing external threads.

DIE PLATE – Tool in the form of a flat plate of good quality cast steel having a number of drilled and tapped holes, employed in hand cutting screw threads on rods of about 5 mm in diameter or smaller.

DOT PUNCH – Light type of sharply pointed centre punch used for light centre dotting along a scribed line to make it more easily visible or to ensure a permanent record of its location.

DOUBLE CUT FILE – A file having teeth formed by two intersecting series of parallel chisel cuts.

DREADNOUGHT FILE – A file with coarsely pitched curved teeth or chisel form produced by cutting curved tooth spaces in the blank.

DRESSER – Tool used in sharpening grinding wheels of abrasive material, usually a diamond tool.

DRILL – Tool used for the production of round holes in metal and other solid material.

DRILL POST AND RATCHET – Portable hand operated drilling machine, comprising three parts—the post or pillar, the arm, and the ratchet brace.

EMERY WHEEL – A grinding wheel made from emery or natural abrasive.

END MILL – A milling cutter having a straight or tapered shank for mounting into a holder for driving. The cutting portion has teeth on the end as well as on the circumference.

EXPANSION REAMER – A type of hand or machine reamer in which the diameter may be slightly increased by an expanding screw.

EXY OUT – A tool for removing a broken stud or bolt from a hole. A hole is drilled into or through the portion of the stud remaining in the hole. Then the exy out is screwed into the hole and when the tool is turned counter clockwise, the broken portion is removed from the hole.

FACE MILL – A milling cutter similar to an end mill but of a larger size.

FEELER GAUGE – Leaf gauge, rather like a multibladed pen knife, the various leaves or blades all having different thicknesses.

FILE – Steel tool used for smoothing rough surfaces, and for reducing the thickness of material by abrasion. The teeth may be single cut or double cut.

FILE CARD – Device fashioned like a wire brush, used to clean dirt and chips from the teeth of a file.

FINISHING TOOL – Tool for taking a light finishing cut, a knife edge tool used with a fine feed or a round nose finishing tool with a coarse feed.

FLAT CHISEL – Commonest form of cold chisel having a long single cutting edge.

FLAT DRILL – Type of drill in general use before the introduction of the twist drill, having flat sides and, therefore, no true cutting action, the flat drill is inclined to have a scraping action rather than cutting.

FLAT FILE – General utility file having a rectangular section, parallel in both width and thickness for about two thirds its length and slightly tapered off in both width and thickness towards the point for the rest of its length. They are double cut on their wide working faces, but are single cut on both their narrow edges.

FLAT ENDED DRILL – by means of which a hole previously drilled to the correct depth with a twist drill may be squared out to a flat bottom, the small "pip" is necessary to ensure that the drill runs concentrically.

FLUTED CHUCKING REAMER – A machine reamer which has straight or helical flutes to provide cutting edges over the entire length of the flutes. Intended for removing a small amount of metal and for finishing a hole accurately and smoothly.

FLY CUTTER – A single point tool for use on milling machines.

FOLLOWON TOOL – Multistage press tool incorporating such operations as blanking, piercing, marking and bending and used in either in a hand or single action power press.

FORMING TOOL – A cutting tool used for forming regular or irregular shapes. The cutting tool is ground to the shape desired and reproduces this shape on the workpiece.

FORM MILL – Milling cutter used for a specific form or shape.

FULLER – Tool used by the smith for making rivetted joints or shaped work on the anvil, or working under the power hammer.

GATE KNIFE – Moulder's tool used for cutting away sand to form the gate passage through which metal must run from the main supply to the mould cavity.

GEAR CUTTERS – Accurately formed cutting tools of hardened steel having shaped teeth that cut the spaces between the teeth of a gear to the precise shape and size required.

GRINDING WHEEL – Abrasive wheels in which the abrasive material is held together by a bonding material. The elements that make up the wheel are (1) abrasive, (2) grain size, (3) bond, (4) grade and (5) structure. There are nine standard shapes and many sizes and grades for grinding practically all kinds of materials.

HACKSAW – Tool employed in cutting bars, rods, tubes etc., of small size, it consists essentially of a frame, a handle and a blade.

HALF ROUND FILE – File which has a section like a segment of a circle (less than a semicircle).

HAMMER (pneumatic) – Hammer held and applied to the work by hand, but actuated by compressed air.

HAMMER (sledge) – Tool with a steel head and a wooden handle, a larger type of hammer which may have the head weighing from, say 2 to 7 kg with the handle varying from, say 60 to 90cm in length.

HAND CHASERS – are made in pairs one for internal work, and the other for external work. These are used only for small threads on brass, or for rounding off the tops and bottoms of threads previously formed by the vee tool in the screw cutting lathe.

HAND FILE – Commonly used file parallel in width from the tang to the tip, but slightly thinned off in thickness for about one third of its length towards the point.

HAND HAMMER – Type of hammer employed by fitters, machinists and so on, the head weighing from 12 to 24 oz.

HAND TAP – A hardened and tempered steel tool for cutting internal threads. It has a thread cut on it, and is fluted to provide cutting edges. A square at the shank end makes it possible to attach a wrench for turning by hand.

HAND TOOL – A general term that applies to wide variety of small tools used by hand instead of being power driven, as layout tools, wrenches and hammers.

HOLLOW MILL – Type of milling cutter in which the work passes through the center of the cutter.

HOB – Fluted rotary cutter used to produce spur, helical, and worm gears, a worm shaped cutting tool having a number of flutes or gashes running across the threads so that a series of cutting edges is formed.

INDENTER – Tool used to form an indentation *i.e.,* a small recess in the surface of an article.

INSERTED TIP TOOL – Lathe tool with a specially hard tip inserted in a certain steel holder.

INSERTED TOOTH CUTTER – Milling cutter with separate teeth fitted into a body or hob.

INTER LOCKING CUTTERS – Milling cutters consisting of two sections. Mating sections are similar to half side cutters or staggered tooth side cutters with uniform or alternate helical teeth so designed that the paths of teeth overlap when in proper assembly.

KEYWAY MILL – Cutter for milling of keyways in shafts, made in standard widths from 1/6 to 3/8 inch and diameters from 1/2 to $1\frac{1}{2}$ inch.

KNIFE FILE – File in transverse section tapers symmetrically from one edge to the other, like a wedge or a knife, used for filing in narrow spaces and cleaning sharp corners.

KNURLING TOOL – Tool to cut a knurl or diamond shaped impression on the surface of a component.

LAP – A tool for finishing internal and external surfaces of the workpiece by charging the lap with a fine abrasive. Always made from metals which are softer than the metal being lapped.

LAPPING STONE – Stone used for lapping operations, especially on tools and cutters of high speed steel and tungsten carbide.

LEAD HAMMER – A soft hammer made of lead for tapping, seating and adjusting work pieces to prevent damaging the finished surface.

LEFT HAND CUTTER – In lathe work, a cutting tool that cuts when fed from left to right or towards the tailstock. For milling cutters, when viewed from the spindle or shank end, the cutter would turn counter-clockwise.

LEFT HAND SCREW – One that screws into the mating part or advances when turned to the left or counterclockwise.

LOADED WHEEL – A grinding wheel that has become dull or filled up from the abrasive particles of the material being ground.

MACHINE TAPS – Taps designed for continuous rather than intermittent cutting and which have less land, more chip space, spiral or straight flutes, and longer shanks on some types.

METAL BAND SAW – A power driven precision saw having an endless blade which runs over two wheels mounted one above the other and used to cut metal.

METAL SLITTING SAW – A thin milling cutter for slotting or cutting off stock in a milling machine.

MILLING CUTTER – Rotary cutter having teeth and spaces on the circumference and side or end faces of steel discs, used on milling machines.

NAIL PUNCH – Punch used by wood workers to drive the head of a nail a little below a surface.

NECKING TOOL – Tool designed and used to reduce the diameter of a component and thus form a neck.

NEEDLE FILE – Small, light files usually restricted in length to 3 to 4 inch, made with a variety of cross-sectional shapes.

NUMBER SIZE DRILLS – A system of drill sizes for small drills in which the diameters are designated by numbers from No 1 the largest, to No 80 the smallest.

OIL STONE – Stone employed for sharpening edged tools.

PARALLEL ROUND FILE – Single cut file of circular cross-section parallel throughout its length, usually known as a blunt round or gulleting file.

PARALLEL SQUARE FILE – Type of file of square section parallel throughout its length, usually known as a square blunt file.

PIERCING SAW – resembling a wood-work fret saw is used for cutting internal holes, slots and shapes, its lower jaw is adjustable for blades of varying lengths.

PIN PUNCH – Type of punch, sometimes called a drive pin punch, employed in removing dowel pins, cotter pins, rivets and similar light objects used as temporary fastenings, also for punching small holes in sheet metal.

PLAIN MILLING CUTTER – A milling cutter that has cutting teeth on the circumference surface only.

PLANER TOOL – Tool used in a planing machine made of different sizes, shapes and materials.

PNEUMATIC TOOL – Design of hand tool operated by air pressure.

PUNCH (hand) – Tool used for producing a rough hole.

PULLEY TAP – A thread cutting tap having a very long shank for tapping set screw holes in pulleys.

RADIUS CUTTER – A side or end milling cutter which has the edges of the teeth ground to specified radius so that it will reproduce the radius on the workpiece.

RAMMER – Tool used by foundry workers for consolidating sand.

RATCHET DRILL – Short type of drill for use in a ratchet brace.

REAMER – Multiedged cutting tool having cutting edges arranged down the length of the tool, the teeth (and the flutes) being either straight and parallel to the axis or alternatively helical with a lead opposite to the direction of rotation.

RECESSING TOOL – Type of tool employed to cut recesses in bored holes.

RIFFLER – File with a curved working surface specially made for filing concave surfaces, such as the insides of oddly shaped holes in castings.

RIGBY HAMMER – Steam hammer used for a wide range of forging work.

RIGHT CUT TOOL – A single point lathe tool, which when used in a lathe, the cutting edge is on the left side and cuts when fed from right to left.

RIGHT HAND CUTTER – A term used to describe both rotation and helix of milling cutters. A cutter that rotates clockwise when viewed from the spindle end is said to have right hand rotation. A cutter has a right hand helix when the flutes slant downward to the right when viewed from the front or twist clockwise when viewed from the end.

RIVET PUNCH – Tool for piercing holes for rivets, few have a spiral cutting edge.

RIVET SET – Tool used to shape the head of a rivet, and sometimes known as cup tool and snap.

ROLLER, KNURLING – Grooved roller for the purpose of producing a knurled or roughened surface on the outside of the work.

ROSE CHUCKING REAMER – Type of reamer with straight flutes to enlarge deep holes or holes of large diameter, in machined parts.

ROSE REAMER – A machine reamer designed so that all the cutting is done on the bevelled ends of the teeth instead of on the sides, sometimes also called ROSE SHELL REAMER.

ROUGH FILE – Type of file having coarsely cut teeth, approximately 20 teeth per inch.

ROUGHING TOOL – Lathe tool designed to take a heavy cut for preliminary roughing cut.

ROUND FILE – Type of file of circular section throughout its length, made in two forms, taper (or slim) and parallel (blunt) .

ROUND NOSE CHISEL – Hand cutting tool used for incising grooves in bearings and bushes, roughing out filleted corners and so on.

ROUND NOSE TOOL – A single point cutting tool having a round nose and with or without side rake angles. With no side rake it may be used to cut either right or left hand.

SAFE EDGE FILE – File with one edge left uncut and it is very useful when working against a shoulder.

SCRAPER – Tool used to reduce the high spots on a surface which has previously been smoothed as far as possible by machining or filing. Scraping is hand operation.

SAW SHARPENING FILE – Type of file used for sharpening saw teeth, its shape being dependent on the shape of the teeth (mill file, cross cut saw file, taper saw file, double ended taper saw file).

SCREW PLATE – A set consisting of the more commonly used taps and dies, a tap wrench, and a die stock, cased in a wooden box.

SCREW SLOTTING SAWS – A series of thin saws of various widths, generally used in an automatic screw machine but sometimes in a milling machine, to cut the slots in the heads of screws.

SCREW TAP – Tool used in producing a screw thread in a hole *i.e.*, an internal thread.

SCRIBER – Metal tool, shaped something like a thin pencil, having a sharply pointed end for marking lines on metal surfaces so as to locate the positions of holes, angles, machining boundaries.

SCRIBING BLOCK – Tool comprising a heavy base, a rigid vertical spindle and a sleeve device capable of travelling up and down the spindle and holding a scriber, used primarily on surface plates for the purpose of scribing lines parallel to a level surface or to another line.

SECOND CUT FILE – Type of file having a grade of cut between a bastard file and a smooth file.

SET – The bend to one side of the teeth of a saw. Also, any deformation of metal or other material stressed beyond its elastic limit.

SET FILE – Type of file bent or set to a special shape to make possible the filing of some part of a workpiece otherwise inaccessible.

SHANK – The non-cutting end of a tool which fits into the holding device for driving, as the taper shank on a drill.

SHAPING TOOL – Cutting tool used in a shaping machine, and is designed to withstand a certain amount of shock, not only when starting, but also when they are forced across the surface of the metal.

SHEARS (hand) – Scissors like cutting tool used for hand cutting sheet metal to the required shapes.

SHELL END MILL – Large type of end mill ranging from $1\frac{1}{4}$ to 3 inch diameter. The body of the cutter is slightly less in length than its diameter, and is made with a hole through the centre to fit a standard arbor.

SHELL REAMER – Type of reamer with a central hole which can be mounted on a suitable arbor. The term shell is used to differentiate this type of reamer from the solid reamer.

SINGLE CUT FILE – A file that has single rows of parallel teeth arranged at an angle across the face.

SINGLE POINT CUTTER – Boring cutter with one point only for use in a boring bar.

SIDE AND FACE MILL – Milling cutter with teeth on the sides and circumference.

SIDE FACE TOOL – Knife edged tool for side turning or facing.

SIDE TOOL – Knife tool used for ordinary turning in a lathe.

SLABBING CUTTER – A wide face milling cutter designed for heavy roughing cuts.

SLIDE BEVEL – Angles between adjacent surfaces of an object, which are not 90 degree are tested with the slide bevel.

SLITTING SAW – Circular saw for use on milling machines to make slitting or cutting through metal, or the cutting of deep narrow slots.

SLOTTING DRILL – Short, straight flute drill made without a point. It is more of a milling cutter than a drill.

SMOOTH FILE – Type of file having a grade of cut between a second cut and a dead smooth. The number of teeth per inch varies considerably for different makes, sizes and shapes, but a smooth file 12" long has about 50 to 60 teeth/inch.

SOFT HAMMER – A hammer, the head of which is made of copper, lead, raw hide or plastic to avoid any damage to the finished surfaces of work-pieces or machine parts.

SOLDERING IRON – Tool used during a soldering operation to heat the solder and the parts to be joined.

SPACING DEVICE (for centre punch) – Simple device which can be fitted to an automatic centre punch to enable light punch marks or dots to be spaced out at regular distances along a scribed line.

SPANNER – Lever tool, also called a wrench, employed for tightening and loosening nuts, bolts, screws etc.

SPANNER WRENCH – A type of wrench having a hook or equipped with pins for tightening or loosening threaded circular collars which have either slots or holes to receive the hook or pins on the wrench.

SPIRAL MILL – Milling cutter with the cutting teeth in the form of a spiral or helix round its circumference.

SPOT FACING TOOL – is similar to the counterbore and is used primarily for producing a shallow, annular, finished surface to mate with a locking washer or the head of a bolt or nut.

SQUARE – The name of tools of various designs and types used for laying out, inspecting, and testing the squareness of workpieces. Also, lines or surfaces which meet at right angles.

SQUARE FILE – Type of file with square section throughout its length, made in two forms taper (or slim) and parallel (or blunt).

SQUARE NOSE TOOL – A single point end cutting tool usually ground with the end cutting edge at approximately 90° to the axis of the point. Used for chamfering or for rough cuts on flat surfaces where sharp corners are necessary. Also, used for finishing cuts on cast iron.

STAGGER TOOTH CUTTERS – Side milling cutters in which the teeth having alternating helix and the drag ends are eliminated to provide more chip clearance.

STOCK AND DIE – Tool used for cutting external threads on cylindrical bars etc. The stock is the name given to the portion of the tool (hand holder) into which the dies are placed and which enables the dies to be rotated without undue effort. The assembled unit is called die and stock.

STRADDLE MILL – Term used when two or more milling cutters are fitted on the arbor with spacing collars between them.

STRAIGHT REAMER – is a cutting tool used to finish or semifinish the surface of a cylindrical hole.

STRAIGHT SHANK DRILL – Twist drill having a shank that is straight rather than tapered and held in a self centering drill chuck for driving.

SURFACE PLATE – Metal plate with an optically flat surface which is used as a plane of reference.

SWAGE BLOCK – Tool, made either in cast iron or steel and weighing roughly 75 kgs, used for shaping and finishing of forgings of a variety of cross sections.

SWAN NECK TOOL – Cranked tool for finishing slender work in a lathe.

SWISS FILE – Class of fine file used on light accurate work.

TANG – Two opposite flats milled at the end of taper shanks which enter a slot in the spindle to prevent slipping and provide a means of removing the shank from the spindle. Also the name of the part of a file that fits into the handle.

TAP – A hardened and tempered steel tool for cutting internal threads which has flutes lengthwise to provide cutting edges for the threads and a square at the end of the shank for turning the tap with a wrench.

TAP DRILL – A twist drill, the size of which will produce a hole that can be tapped with a specified tap to give a thread of required depth or percentage of full depth.

TAP EXTRACTOR – A tool or device for removing broken taps from holes. Prongs extend down into the flutes of the broken portion. By attaching a wrench to the extractor, the broken part in the hole may be loosened and backed out.

TAP GROOVING CUTTER – Relieved milling cutter having a contour specially designed for cutting the grooves or flutes in a tap.

TAPER PIN REAMER – Reamer employed to open out a drilled Ahole so that a standard taper pin can be fitted.

TAPER TAP – is used to machine an internal taper thread such as a pipe thread. It is chamfered at its small end for a distance equal to 4 or 5 times the distance between threads.

TAP WRENCH – Tool which grips the square ended shank of a tap and enables it to be worked through a hole in hand tapping.

TEE SLOT MILL – Milling cutter used for cutting the cross groove of a Tee slot.

TOOL BIT – Small section of high speed steel, suitably ground, and held in a tool holder.

TOOL MAKERS FILE – File which in width and thickness is smaller than an ordinary file generally obtainable in sizes from 2 to 12 inch in length.

TOOL MAKERS FLAT – Extremely accurate surface plate used by tool makers.

TRAMMEL – Instrument mainly used for marking out arcs of circles too large in diameter to permit the use of dividers.

TREPANNING TOOL – A tool holding device designed to hold a cutting tool bit. This device is rotated in a drilling machine to cut out large circles or holes from solid metal by cutting a narrow groove around the metal.

TRIANGULAR FILE – File having a section shaped like an equilateral triangle, usually known as a three square file.

TRIMMING DIE – Tool used to remove the surplus material from the edge of an article.

TROWEL – Tool most frequently used by moulders and core makers in the execution of their work.

TRY SQUARE – Adjacent surfaces of an object, whether they are at right angles or not is tested with the trysquare.

T SLOT CUTTER – A one piece, T shaped cutter which has a narrow neck connecting the cutter to the shank, the diameter and width of the cutter being standard for each number of T slot to suit bolts of various sizes.

TURNING TOOL – Tool of hardened material, suitably ground, for the purpose of forming cylindrical or tapered work in a lathe.

TWIST DRILL – Drill with spiral flutes around its circumference. These flutes leading from the cutting faces, called lips, provide an exit for the metal that is removed when the hole is being drilled.

TWO LIP END MILL – A type of end milling cutter having two cutting edges or lips in which the flutes may be straight or helical. Used for the fast removal of metal when slots or keyways are to be cut on the milling machine.

UNIVERSAL SQUARE – Tool used to find the centre of a circle without the use of other marking out instruments.

VEE BLOCK AND CLAMP – are used for holding circular bars for centering, drilling etc.

VIXEN FILE – A flat file with curved teeth which is largely used for filing soft metals.

WARDING FILE – Small file, usually about four to six inches long, having a rectangular section, used by locksmiths when filing keys to fit the wards of a lock or for filing the actual wards.

WASHER CUTTER – Tool designed to cut washers, or round discs, with or without holes in their centers, from sheet metal.

WHEEL (grinding) – Power driven wheel consisting of abrasive particles, held together by artificial or natural, mineral, metal or organic bonds.

WHEEL DRESSER – A tool or device used for dressing the face of grinding wheels.

WIRECUTTER – Tool used to cut wire and small diameter rod.

WOODRUFF KEYWAY MILL – Milling cutter for cutting standard woodruff keyways.

CHAPTER 8

INSTRUMENTS AND GAUGES

ACME THREAD GAUGE – A gauge used for checking and testing the 29° angle and the width at the end of a thread cutting tool while grinding it. It is also used for setting the tool square with the axis of the workpiece.

ANGLE PLATES – Some work has to be set at a particular angle to the surface of the table, and this is often effected by bolting it to an adjustable angle plate.

ARDOMETER – A type of total radiation pyrometer.

BORING BAR MICROMETER – On boring operations, it is often necessary to adjust the cutter setting by a few thousandths of an inch. With this, it is possible to determine exactly the depth of cut taken.

CALIPER – Instrument used for measuring distances between or over surfaces, or for comparing dimensions of workpiece with such standards as plug gauges, graduated rules etc.

CENTRE SQUARE – is intended for marking the centres of round or square stock.

CLINOMETER – Instrument used for measuring angles relative to the horizontal plane.

DEPTH GAUGE – Gauge used for the purpose of measuring the distance between two surfaces, to find or check the depth of a counterbored hole, a recess etc.

DIAL GAUGE – Instrument indicating the movement of a projecting plunger by the movement of a hand or pointer over a graduated dial.

DIAL SHEET GAUGE – Gauge fitted with a dial or clock for measuring or comparing the thickness of sheet metal plates.

DIVIDERS – These are essentially scribers for marking arcs, and circles, hence their points are hard and sharp. They are used for dividing distances into a number of equal parts. Also used for transferring dimensions from a rule to the work.

DRILL GAUGE – Tool for measuring the size of drills.

DRILL GRINDING GAUGE – A tool for checking the angle and length of the cutting lips on a twist drill while grinding it. Also it may be called a drill point gauge or a drill angle gauge.

ENGINEERS TAPERS – are intended for measuring hole sizes, slot widths, caliper settings and so on.

FEELER GAUGES – Consist of a number of thin metal blades, each of a different thickness and in the desired number of dimensional steps. They are used for checking clearances.

FIRTHS HARDOMETER – Instrument for measuring the hardness of materials, particularly metals.

GAUGE – A tool or instrument for checking or measuring the sizes of metal parts to determine whether the dimensions are within the specified limits.

GAUGE BLOCKS – are used mostly for reference in setting gauges and for accurate measurement in tool, gauge, and die manufacture.

HEIGHT GAUGE – Marking out or measuring tool designed on the lines of a scribing block, but capable of more accurate duty.

HERMOPHRODITE CALIPER – A caliper having one leg similar to that of an inside caliper, while the other leg is pointed like a divider leg.

HOOK RULE – A steel rule with a projecting piece or hook at one end at right angles to the edge of the rule.

INDICATING INSTRUMENTS – The instruments that provide visual observation of readings, such as pressure gauges, draft gauges, fuel oil meters and thermometers for fuel, feed water and flue gas temperatures etc.

INDICATOR – A sensitive instrument which shows slight variations when testing the trueness, or alignment of a workpiece, fixture or machine part such as a milling vice. Dial indicators may be graduated to read small linear variations.

INSIDE VERNIER CALIPER – is intended mainly for checking internal dimensions, such as the diameter of bores, and so on. Vernier protractor is used for measuring angles accurately.

JO BLOCKS – These refer to Johansson precision gauge blocks.

KNIFE EDGE VERNIER CALIPER – has jaws which are reduced to a very narrow edge. These are useful for measuring in restricted places or on curved surfaces.

LEVEL, HYDROSTATIC – U tube device employed during the levelling and setting of work too long for the use of a spirit level.

LEVEL, SPIRIT – Instrument consisting essentially of a small sealed glass tube containing spirit (alcohol or ether) so mounted that the enclosed bubble of air occupies a central position only when the instrument is placed horizontally.

LIMIT GAUGE – A gauge that represents a limiting (maximum and minimum) size within which the work will be acceptable.

MICROMETER – A precision, screw adjusted measuring instrument with which dimensions can be read in thousandths and ten thousandths of an inch.

MICROMETER COLLAR – A dial on the screw of a machine to indicate the extent of the movement of the screw or parts attached to the screw and usually graduated to read thousandths of an inch.

MICROMETER DEPTH GAUGE – The ordinary depth gauge has been elaborated by the addition of a micrometer head.

MINNIMETER – Delicate form of indicator showing the movement of a projecting plunger usually placed at one end.

ODD LEG – Caliper having one leg bent inwards at its ends, like the two legs on a pair of inside calipers, the other leg being pointed or having a separate hardened point attached to it.

OPTICAL FLAT – Plate generally made of glass or quartz and used as a standard of reference as a flat surface.

OPTICAL PROTRACTOR – Angle measuring instrument fitted with a scale read by an optical magnifying device.

PARALLEL BLOCK – Steel or cast iron block used when setting up work for machining, marking out or measuring.

PIN GAUGES – End measuring pins for determining bore diameters.

PLUG GAUGE – A gauge on which the outside measuring surfaces are designed to test the specified dimensions of holes. May be straight or tapered, plain or threaded, and of any cross-sectional shape.

PLUMB BOB – Weight hung on the end of a cord so that when the latter is supported at its top end the cord will hang vertical.

PNEUMATIC MICROMETER – An instrument designed to measure, among other things, the errors in the size, shape and alignment of bores, with compressed air.

PROTRACTOR – Instrument graduated so as to enable the measurement of angles to be made in degrees.

PYROMETER – Specified type of thermometer used to measure high temperatures in the production and heat treatment of metals and alloys.

RADIUS GAUGE – Tool used for checking the radius of small fillets or rounded corners.

RECORDING INSTRUMENTS – Instruments which provide a permanent record of readings such as steam pressure, steam flow, air flow, flue gas temperature, feed water temperature, fuel flow and fuel temperature.

RING GAUGE – A gauge in which the inside measuring surfaces are circular in form. A ring gauge may be cylindrical or conical, plain or threaded.

SCLEROSCOPE – Instrument for measuring hardness.

SCREW GAUGE – Type of gauge for checking the accuracy of a screw thread.

SCREW PITCH GAUGE – A small tool with a number of blades, each having the same number of notches per inch as the thread it represents. It is used to find the pitch and number of threads per inch on a screw, tap or die.

SCREW THREAD MICROMETER – Type of micrometer with measuring points specially shaped to enable the measurement of pitch diameter and thickness of thread.

SCRIBER – Used for making lines on the work when marking off.

SINE BAR – Tool used for the accurate setting out of angles by arranging to convert angular measurements to linear ones.

SLIP GAUGE – Form of gauge made of hardened steel with two parallel faces worked to high precision and at a known distance apart.

SNAP GAUGE – A type of fixed gauge, plain or adjustable, arranged with inside measuring surfaces for checking diameters, lengths, thicknesses or widths of workpieces.

SPRING CALIPERS – Inside or outside calipers for measuring workpieces in which the tension against the nut is maintained by a circular spring at the end.

STEEL RULE – A thin flat measuring tool graduated in either fractional or decimal part of an inch or cm and made in many styles and lengths.

STRAIGHT EDGE – Strip of metal, usually steel, with one edge straight, used to determine the flatness of a surface.

SURFACE GAUGE – Tool used for marking out and setting up work in conjunction with a surface plate, used also for levelling workpieces when fixing them to the worktables of machine tools *e.g.*, planers, shapers, millers etc.

TAPER PLUG GAUGE – A solid internal gauge in the form of a frustrum of a cone having diameter, taper and length suitable for measuring the dimensions of an internal taper of specified size.

TAPER RING GAUGE – An external gauge, the internal diameter of which confirms to the frustrum of a cone having diameter, taper and length suitable for measuring the dimensions of an external taper of specified size.

TEMPLATE – A flat pattern or guide plate usually made from sheet metal and used as a gauge or guide when laying out, drilling, forming in a machine or filing irregular shapes on metal workpieces.

TELESCOPING GAUGES – Adjustable gauges which consists of a handle attached to a fixed contact within which a plunger contact expands or telescopes under spring tension when the gauge is inserted into the hole that is to be measured.

THREAD GAUGE – Tool for checking the form and pitch of screw threads.

TOOL MAKERS MICROSCOPE – Instrument used in accurate inspection of thread forms, form tools and complex contours.

TORQUE METER – Torsion measuring device fitted to a rotating shaft or component to measure the torque developed at any given moment under actual running conditions.

TORSIOGRAPH – Instrument used to measure and record the amplitude and frequency of torsional vibrations in a rotating shaft or other part.

TRAMMELS – are generally used for laying off and checking dimensions of several feet, they are generally used in conjunction with a large steel rule, which may be from 3 to 10 ft long.

TRANSFER CALIPER – A caliper designed to take measurement in recesses or over a projection. One leg can be separated or opened to remove the caliper and then returned to the measured size.

TRY SQUARE – Instrument used for setting and checking a line, an edge or a face which is required to be at right angles to some other plane which should be regarded as a datum or reference plane.

VACUUM GAUGE – Gauge used to measure the amount of vacuum in any vessel in which a pressure lower than atmospheric pressure is produced by evacuation of air, vapour and gas.

VERNIER – Small-scale, for making accurate measurement, carrying a certain number of graduations equalling in their combined length a different number of graduations, usually one more or one less, on the main scale of the instrument to which the vernier is attached.

VERNIER CALIPER – Precision measuring instrument consisting of a beam or rule having a fixed jaw at one end, and a straight jaw to which is attached a vernier scale.

WATER GAUGE – A gauge that shows the proper water level which must be maintained in a boiler to avoid overheating damage.

WIRE GAUGE – Sizes of standard wire and sheet metal are normally expressed not by a dimension in cm, but by a number or gauge size.

WHITWORTH GAUGE – Type of plug, and caliper gauge introduced by Joseph Whitworth.

WICKMAN GAUGE – Special form of adjustable snap or horse shoe type of gap gauge.

CHAPTER 9

CASTING

AERATOR – A device for fluffing (or decreasing the density of) and cooling the sand by the admixture of air.

AIR GATE – A vertical channel for the removal of gases from the mould; checking of the filling of the mould cavity with metal and feeding up the casting with metal during solidification.

AIR HOLE – Hole in a casting caused by air or gas trapped in the metal during solidification.

AIRLESS BLAST CLEANING – A process whereby the abrasive material is applied to the object being cleaned by centrifugal force generated by a rotating vane type wheel.

ANTIPIPING (material) – Usually refers to an insulating material placed on the top of a sprue or riser that keeps the metal in liquid or semiliquid form for a long period of time and minimizes the formation of the usual conical pipe or shrink in the top of a sprue or riser.

ARBOR – A metal barrel, frame, or plate to support or carry part of a mould or core.

ATMOSPHERIC RISER – Blind riser which employs atmospheric pressure to aid feeding.

BACK DRAFT – Taper or draft which prevents removal of pattern from the mould.

BACKING SAND – Sand between the facing sand and the flask.

BAKED CORE – The core which has been subjected to heating or baking until it is thoroughly dry, as opposed to green sand core which is used in the moist state.

BASIN – A cavity on top of the cope into which molten metal is poured before it enters the sprue.

BEDDED IN MOULD – is the mould, the bottom half of which is made in the sand in the floor of the foundry. It may be covered with a cope, or cast open, according to the type of work.

BINDER – Material used to hold the grains of sand together in moulds or cores. May be cereal, oil, clay, resin, pitch etc.

BINDER, PLASTIC (resin) – Synthetic resin material used to hold grains of sand together in moulds or cores, may be phenol formaldehyde or urea formaldehyde thermosetting types.

BLACKING – Carbonaceous material for coating mould or core surfaces.

BLACK LEAD – Graphite for facing moulds and cores.

BLAST CLEANING – Removal of sand or oxide scale from castings by the impinging action of sand, metal shot or grit projected under air, water or centrifugal pressure.

BLEED – Molten metal oozing out of a casting stripped or removed from the mould before solidification.

BLENDED SAND – Mixture of sands of different grain sizes, clay content etc. to produce one, possessing characteristics more suitable for foundry use.

BLIND RISER – An internal riser which does not reach to the exterior of the mould.

BLISTER – Defect on the surface of a casting appearing as a shallow blow with a thin film of metal over it.

BLOWN CASTINGS – Castings in which bubbles, or blowholes, have been caused through gases, steam etc., generated when the mould is cast, finding their way into the metal.

BOND CLAY – Any clay suitable for use as a bonding material in the moulding sand.

BORIC ACID – Inhibitor used in facing sand for magnesium base and aluminium base alloys high in magnesium to prevent reaction with moisture in the sand.

BORON TRICHLORIDE – A product used for degasification of aluminium alloys.

BOSSES – Bosses are often located on a wall of a casting and should be so designed that a heavy section of metal leads to the riser.

BOT – Clay wedge used in a cupola to stop the hole through which the metal is run.

BUCKLE – Defect in a casting surface appearing as an indentation resulting from an expansion scab.

BURN ON – Adhesion of sand to the casting, usually due to the metal penetrating into the sand.

BURN OUT – Usually refers to the removal of the disposable wax or plastic pattern in the investment moulding process by heating the mould gradually to a sufficiently high temperature to consume any carbonaceous residues.

BUTT RAMMAR – The flat end of the molders rammer.

CALCIUM BORIDE – An alloy of calcium and boron corresponding (when pure) to the formula CaB_6 containing about 61% boron and 39% calcium, and used in deoxidation and degasification of nonferrous metals and alloys.

CALCIUM MANGANESE SILICON – An alloy containing 17 to 19% calcium, 8 to 10% manganese, 55 to 60% silicon and 10 to 14% iron used as a scavenger for oxides, gases and nonmetallic impurities in steel.

CALCIUM SILICON – An alloy of calcium, silicon and iron containing 28 to 30% calcium, 60-65% silicon and 6% max iron, used as a deoxidizer and degasifier for steel and cast iron.

CASTING – Metal poured into a mould to form an object. Act of pouring molten metal into a mould.

CASTING STRAINS – Strains resulting from internal stresses created during cooling of a casting.

CAVITY, MOULD or DIE – Impression or impressions in a mould or die that give the casting its shape.

CENTRIFUGAL CASTING – Process of filling moulds by pouring the metal into a sand or metal mould revolving about either its horizontal or vertical axis and continue pouring the metal into the mould that is being revolved before solidification of metal is complete. Molten metal is moved from the center to the periphery by centrifugal action.

CERAMIC MOULD – Mould in which the refractory and binder are such that when fired at high temperature, a rigid structure is formed. The mould can be made in a flask or in the form of a shell.

CEREAL – Substance derived from corn flour, which is added to core and moulding sands to improve their properties for casting production.

CHALK TEST – Method of crack detection which consists of applying a penetrating liquid to the excess from the surface which is then coated with whiting or chalk. After a short time, the penetrant seeps out of the cracks into the whiting, causing an appreciable difference in whiteness.

CHAMOTTE – Coarsely graded refractory material prepared from calcined clay and ground firebrick mulled with raw clay, used in steel foundries.

CHAPLET – A metallic insert or support to hold the core in position in the mould.

CHEEK – Intermediate sections of a flask inserted between cope and drag.

CHILL – A metal object placed on the outside or inside a mould cavity to induce more rapid cooling at that point and thereby produce hard zone *i.e.*, hard, unmachinable surface.

CHILL TEST – Method of determining the suitability of a gray iron for specific castings through its chilling tendency, as measured from the tip of a wedge shaped test bar.

CHILLED IRON – Cast iron poured against a chill to produce a hard unmachinable surface.

CHOKE – Restriction in a gating system to control the flow of metal beyond that point.

CHVORINOV'S RULE – Solidification time is proportional to the square of the volume of the metal and inversely proportional to the square of the surface area.

CLEANING – Process of removing sand, surface blemishes etc. from the exterior and interior surfaces of a casting. Includes degating, tumbling, or abrasive blasting, grinding off gate stubs.

COD – A sand projection left behind in the mould by some patterns. Strictly speaking it is a core.

COLD SHUT – Where two streams of metal do not unite thoroughly in a casting.

COMBINATION DIE – A die casting die having two or more cavities of dissimilar parts.

CONTRACTION – Act or process of a casting becoming smaller in volume and/or dimensions during the solidification of the metal or alloy which composes the casting.

COPE – The upper or top most section of a flask, mould or pattern.

COPE **(false)** – A temporary cope which is used only to establish the parting line.

CORE – Separable part of the mould, usually made of sand and generally baked, to create openings and various shaped cavities in the castings. Also used to designate the interior portion of an iron base alloy which after case hardening is substantially softer than the surface layer or case.

CORE BINDER – Any material used to hold the grains of core sand together.

CORE BOX – Box with an opening in which the core is formed.

CORE PRINT – An extension of the pattern for locating the core or an extension of the mould cavity for locating the core.

CORE **(ram up)** – Core attached to the pattern and rammed up in the mould, where it remains when the pattern is withdrawn.

CORE SHIFT – Defect resulting from movement of the core from its proper position in the mould cavity.

CORE VENTS – A wax product, round or oval in form, used to form the vent passage in the core.

CRUSH – Casting defect appearing as an indentation in the surface due to displacement of sand in the mould, usually at the joint surfaces.

CUTS – Defects in castings resulting from erosion of the sand by the molten metal pouring over the mould or core surface.

DEGASSIFIER – A material employed for removing gases from metals and alloys.

DEOXIDIZER – A material used to remove oxygen or oxides from metals and alloys.

DESULPHURIZER – A material used to remove sulphur from molten metals and alloys.

DIE CASTING – Pouring molten metal under pressure into metal moulds.

DIRECTIONAL SOLIDIFICATION – It refers to the solidification which proceeds along a cast member in the direction of the hotter metal.

DIRT – Any extraneous material entering a mould cavity and usually forming a blemish on the casting surface.

DRAFT – Taper allowed on the vertical faces of a pattern to permit removal of it from the sand mould without excessive rapping or tearing of the mould walls.

DRAG – The lower or bottom section of a mould or pattern.

DRY SAND MOULD – A mould made of prepared moulding sand dried thoroughly before being filled with metal.

EROSION SCAB – Casting defect occurring where the metal has been agitated, boiled, or has partially eroded away the sand, leaving a solid mass of sand and metal at that particular spot.

EXPENDABLE PATTERN – In investment moulding, the wax or plastic pattern that is left in the mould and later melted and burned out.

EXPANSION SCABS – Rough thin layers of metal partially separated from the body of the casting by a thin layer of sand and held in place by a thin vein of metal.

EXTERNAL CHILLS – Various materials of high heat capacity such as metals, graphite etc. forming parts of walls of the mould cavity to promote rapid heat extraction from molten metal.

EXTERNAL PRESSURE CASTINGS – In this process, highly fluid metal is forced under considerable external pressure into metal moulds, the pressure being maintained until solidification is complete.

FACING SAND – Specially prepared sand in the mould adjacent to the pattern to produce a smooth casting surface.

FALSE CHEEK – A cheek used in making a three part mould in a two part mould.

FALSE ODDSIDE – Permanent oddside made of plaster or other material.

FEED HEAD – A reservoir of molten metal provided to compensate for contraction of metal as it solidifies, by the feeding down of liquid metal to prevent voids. Also called a RISER.

FILLETS – Properly positioned fillets materially increase the strength and soundness of the castings. They reduce shrinkage cracks and erosion of sand at sharp intersections.

FIN – A thin piece of metal projecting from a casting at the parting line or at the junction of the cores or of cores and mould etc.

FLASH – Thin fin or web of metal extending from the casting along the joint line as a result of poor contact between cope and drag moulds.

FLASK – Container in which a mould is made.

FLOW OFF GATE – Channel cut from the mould to the riser.

FLUORESCENT CRACK DETECTION – Application of fluorescent liquid to a part, then removing the excess from the surface, which is then exposed to ultraviolet light. Cracks show up as fluorescent lines.

GAS HOLES – Rounded cavities caused by generation or accumulation of gas or entrapped air in a casting, holes may be spherical, flattened or elongated.

GATE – The location where the molten metal enters the casting cavity.

GATED PATTERN – One or more patterns with gating systems attached.

GATING SYSTEM – Combination of channels, cavities and other elements of a casting mould which are intended ·to feed molten metal into a mould, fill in the latter, and feed up the casting with metal during solidification.

GRAVITY DIE CASTING – is one in which the fluid metal is poured by hand into the metal moulds and around metal cores. The only pressure applied is that exerted by the head of metal in the pouring gate.

GREEN SAND – Prepared moulding sand in the moist or as mixed condition.

GREEN SAND MOULD – Mould made and cast in damp sand.

GUIDE PIN – The pin used to locate the cope in the proper place on the drag.

HORN GATE – A curved gate in the shape of a horn arranged to permit entry of molten metal at the bottom of the casting cavity.

HOT TEARS – Cracks in castings formed at elevated temperatures, usually by contraction stresses.

IMPRESSION – Cavity in a die casting die or in a mould.

INCLUSIONS – Particles of slag, sand or other impurities such as oxides, sulphides, silicates etc., trapped mechanically during solidification or formed by subsequent reaction of the solid metal.

IN GATE – Channel out from the bottom of the runner into the mould. It is used in cases where the runner does not enter the mould direct.

INJECTION – Forcing molten metal into a die casting die.

INOCULATION – Process of adding some material to molten metal in the ladle for the purpose of controlling the structure to an extent not possible by control of chemical analysis and other normal variables.

INSULATING SLEEVE – Hollow cylinders or sleeves formed of gypsum, diatomaceous earth, pearlite, vermiculite etc., placed in the mould at sprue and riser locations to decrease heat loss and rate of solidification of the metal contained in them.

INTERNAL CHILLS – Solid pieces of metal or alloy, similar in composition to the casting, placed in the mould prior to filling it with molten metal.

INVESTMENT MOULDING – Moulding using a pattern of wax, plastic or other material which is invested or surrounded by a moulding medium in slurry or liquid form. After the moulding medium has solidified, the pattern is removed by heating the mould, leaving a cavity for reception of molten metal. Also called LOST WAX PROCESS or PRECISSION MOULDING.

LADLE – Metal receptacle lined with refractory for transportation of molten metal.

LOAM – A course, strongly bonded moulding sand used for loam and dry sand moulding.

LOAM MOULDING – A system of moulding especially for large castings, wherein the supporting structure is constructed of brick. Coatings of loam are applied to form the mould face.

MAGNETIC CRACK DETECTION – Method of locating cracks in materials which can be magnetized, done by applying magnetizing force and applying finely divided iron powder which then collects in the region of the crack.

MASTER PATTERN – The pattern from which the working pattern is cast.

MATCH – A form of wood, plaster of paris, sand, or other material on which an irregular pattern is laid or supported while the drag is being rammed.

MATCH PLATE – A metal or other plate on which patterns, split along the parting line, are mounted back to back with the gating system to form an integral piece.

METAL PENETRATION – Defect in the casting surface which appears as if the metal has filled the voids between the sand grains without displacing them.

MOULD CAVITY – Impression left in the sand mould by the pattern. Also called MOULD.

MOULD CLAMPS – Devices used to lock or hold cope and drag together.

MOULD HARDENER – In sand moulds in which sodium silicate is the binder, injection of CO_2 causes a chemical reaction which results in a rigid structure.

MOULD WASH – Usually an aqueous emulsion containing various organic or inorganic compounds or both, which is used to coat the face of a mould cavity. Materials include graphite, silica flour etc.

MOULDING SAND – Mixture of sand and clay suitable for mould making.

MOULD WEIGHTS – Weights placed on top of moulds to offset internal and ferrostatic pressure.

MOULDABILITY – The ability of the sand mixture to fill in perfectly the cavity of a flask with a pattern or a corebox.

MULTIPLE MOULD – Composite mould made up of stacked sections. Each section incorporates a complete gate for casting. All castings are poured from a central downgate.

ODDSIDE – Support used for supporting a pattern whilst the drag is being rammed up.

OPEN SAND CASTING – A casting poured in a mould which has no cope or other covering.

PARTING COMPOUND – Material dusted or sprayed on a pattern or mould to prevent adherence of sand.

PATTERN – Model of wood, metal, plaster or other material used in making a mould.

PATTERN ALLOWANCES – The dimensions on the pattern differ from those on the drawing to allow for metal shrinkage, distortion, core shift, draft and machine finish. Such allowances are made by the pattern-maker.

PEEN – Small end of a moulders rammer.

PERMANENT MOULD – A long life mould into which metal is poured by gravity.

PILOT CASTING – Usually the first casting made from a production pattern and examined for dimensional accuracy, quality and other features before the pattern is placed on the line.

PIT MOULD – Mould in which the lower portions are made in a suitable pit or excavation in a foundry floor.

PLASTER MOULDING – Gypsum or plaster of paris is mixed with fibrous talc, with or without sand, and with water to form a slurry, that is poured around a pattern. In a short period of time, the mass sets or hardens sufficiently to permit removal of the pattern. The mould so formed is baked at elevated temperature to remove, all moisture prior to use.

PLASTICITY – The ability of the sand mixture to acquire the outlines of a pattern or corebox under the action of external forces and retain the shape acquired without destruction.

PLASTIC PATTERN – Pattern made from any of the several thermosetting type synthetic resins such as phenol formaldehyde, epoxy etc. Small patterns may be cast solid, but large ones are usually produced by laminating with glass cloth.

POURED SHORT – Casting which lacks completeness due to the cavity not being filled with molten metal.

POURING CUP – Part of the mould that receives the molten metal from the pouring ladle and transfers it further into the cavity of the mould.

PRINT – Wooden projection put on to a pattern to provide supports for the cores in a mould.

PROGRESSIVE SOLIDIFICATION – is the freezing of the metal from the mould metal inferface toward the center of the cavity.

PUSHUP – An indentation in the casting surface due to displacement (expansion) of the sand in the mould.

RAMMER – Tool used in a foundry for ramming the sand.

RAPPING – Tapping of the pattern with a mallet in order to loosen it as it is drawn from the mould.

RAPPING BAR – A pointed bar (or rod) made of steel or other metal, which is inserted vertically into a hole in a pattern or driven into it, then struck with a hammer on alternate sides to cause vibration and loosening of the pattern from the sand.

RELIEF SPRUE – The second sprue at opposite end of the runner to relieve pressure created during pouring operation.

RIBS – are used primarily as stiffeners and reinforcing members. If properly designed and located in difficult castings, serve to check the possibility of hot tears or cracks during solidification.

RISER – Reservoir of molten metal attached to the casting to compensate for the internal contraction of the casting during solidification.

RUNNER – The portion of the gate assembly which connects the downgate or sprue with the casting.

SAG – Defect which appears as an increase or decrease in metal section due to sinking of sand in the cope (decreased section) or sagging in the core (increased section).

SAND BLAST – Sand driven by a blast of compressed air (or steam). Used to clean castings, to cut, polish or decorate glass and other hard substances.

SCAB – A blemish on the casting caused by eruption of gas from the mould face.

SEAM – Surface defect on a casting similar to a cold shut, but not as severe.

SEMICENTRIFUGAL CASTING – is one in which the mould, usually a stacking of several flat sand or die moulds, is rotated about a vertical axis.

SHOT – Abrasive blast cleaning material.

SHRINKAGE ALLOWANCES – are provided to take care of the contraction in dimensions of the solidified casting as it cools from its freezing temperature to room temperature.

SHRINK HOLE – A hole or cavity in a casting resulting from contraction and insufficient feed metal, and formed during the period the metal changed from the liquid to the solid state.

SHELL MOULDING – Process in which clay free silica sand coated with a thermosetting resin or mixed with the resin is placed on a heated metal pattern for a short period of time to form a partially hardened shell.

The unaffected sand mixture is removed for further use. The pattern and the shell are then heated further to harden or polymerize the resin sand mix, and the shell is removed from the pattern.

SKIMMING GATE – Part of the mould that retains non-metallic inclusions or impurities and delivers sound metal further into the runners.

SLEEK – Term meaning to make smooth. It is applied to the trowelling of a sand surface.

SLUSH CASTING – Casting made by pouring an alloy into a metal mould, allowing it to remain sufficiently long to form a thin solid shell, and then pouring out the remaining metal.

SNAG – Removal of fins and rough places on a casting by means of grinding.

SNAP FLASK – Moulding box, hinged on one side so that it may be opened to allow the finished mould to be removed.

SPRIGS – Small pegs of wood or metal used to strengthen weak portions of a mould or to assist in the mending up of a damaged mould.

SPRUE – The vertical portion of the gating system where the molten metal first enters the mould.

STACK MOULDING – Moulding method in which the half mould forms the cope and drag. They are placed on top of the other and moulds stacked one over the other are poured through a common sprue.

STRICKLE – Piece of wood by means of which surplus sand is removed from a moulding box or other surface. They may also be used to shape sand surfaces into any required shape.

SURVIVABILITY – The ability of a moulding mixture to retain its initial properties for a specified length of time.

TRUE CENTRIFUGAL CASTING – is used for producing thin or thick walled hollow cylinders with a bore concentric with the outside.

VENT – Channel made in the sand in the vicinity of a mould to allow steam, gases etc., generated when sand and molten metal come into contact with one another, to escape.

VENT ROD – A piece of wire or rod to form the vents *i.e.*, holes in the sand, for the escape of gases.

VENT WAX – Wax in rod shape placed in the core during manufacture. In the oven the wax is melted out leaving a vent or passage.

WASH – Casting defect resulting from erosion of sand. Also refers coating materials applied to moulds, cores etc.

WASTER – Faulty casing.

WHIRL GATE – Gating system in which the metal enters a circular reservoir at a tangent and so whirls around, leaving dirt and slag behind before passing into the mould cavity.

WHISTLERS – Small openings from isolated mould cavities to allow gases to escape easily.

YIELDABILITY – The ability of the sand mixture to decrease in volume so as not to limit the contraction of a solidifying casting.

CHAPTER 10

WELDING

ARC – A sustained electric discharge, where current flows through the gap between two electrodes.

ARC CUTTING – Process which melts the metals to cut with the heat of an arc between an electrode and the base metal.

ARC EYE – A burn on the exterior surface of the operators eye, due to its exposure to an open arc. Also called FLASH EYE.

ARC GAP – Distance between the tips of two electrodes, normally between an electrode and the workpiece. Also known as ARC LENGTH.

ARC PLASMA – A gas that has been heated to an at least partially ionized condition, enabling it to conduct an electric current.

ARC SPOT WELD – Spot welding made by an arc welding process.

ARC WELDING – A group of welding processes which produces coalescence of metals by heating them with an arc, with or without the application of pressure, and with or without the use of filler metal.

ATOMIC HYDROGEN WELDING – An arc welding process which produces coalescence of metals by heating them with an electric arc maintained between electrodes in an atmosphere of hydrogen. Shielding is provided by hydrogen.

AUTOGENOUS WELD – A fusion weld made without the addition of filler metal.

AUTOMATIC OXYGEN CUTTING – Cutting with an equipment without constant observation and adjustment of the controls by an operator.

AUTOMATIC WELDING – Welding which permits the operation without adjustment of controls by an operator.

BACKHAND WELDING – A welding technique in which the welding torch or gun is directed opposite to the progress of welding.

BACK WELD – A weld deposited at the back of a single groove weld.

BARE ELECTRODE – A filler metal electrode of a single metal or alloy, produced into a wire, strip or bar form and without any coating or covering on it.

BARE METAL ARC WELDING – Process which produces coalescence of metals by heating them with an electric arc between a bare or lightly coated metal electrode and the work piece.

BASE METAL – The metal to be welded, brazed, soldered or cut.

BLIND JOINT – A joint, no portion of which is visible.

BORAX – is the old standard flux for brazing, exists in two forms—ordinary borax and amorphous or fused borax.

BRAZING – A group of welding processes that produces coalescence of materials by heating them to the brazing temperature, using a filler metal having a liquidus above 450°C and below the solidus of the base metal.

BUTT JOINT – A joint between two members aligned approximately in the same plane.

CARBON ARC CUTTING – Cutting of base metals by melting them with the heat of an arc produced between a carbon electrode and the base metal.

CARBON ELECTRODE – A non-filler metal electrode used in arc welding or cutting, consisting of a carbon or graphite rod, which may be coated with copper or other coatings.

CARBON ARC WELDING – A brazing process that produces coalescence of metals by heating them with an electric arc produced between two carbon electrodes. The filler material is distributed in the joint by capillary action.

COLD WELDING – A solid state welding process in which pressure is applied at room temperature to produce coalescence of metals with substantial deformation at the weld.

COMPOSITE ELECTRODE – Multicomponent filler metal electrodes in various physical forms such as standard wires, tubes and covered wire.

CONTINUOUS WELD – A weld that extends continuously from one end of a joint to the other.

COVERED ELECTRODE – An electrode consisting of a core of a bare electrode or metal cored electrode to which a covering, sufficient to provide a slag layer on the weld metal, has been applied.

COVER PLATE (**eye protection**) – A removable pane of colourless glass, plastic coated glass or plastic that covers the filter plate and protects it from weld spatter, pitting or scratching when used in a helmet, hood or goggles.

CRATER – A depression at the termination of a weld bead.

DECARBURIZING FLAME – A flame which removes carbon from the molten metal.

DEPOSITION EFFICIENCY (**arc welding**) – Ratio of the weight of deposited metal to the net weight of the filler metal consumed, exclusive of stubs *i.e.*, left out electrode bits.

DEPTH OF FUSION – Distance that fusion extends into the base metal or previous pass from the surface melted during welding.

DIFFUSION WELDING – A solid state welding process that produces coalescence of the faying surfaces by the application of pressure at elevated temperature.

EDGE JOINT – A joint between the edges of two or more parallel or nearly parallel members.

ELECTROGAS WELDING – Arc welding in which coalescence is effected by heating the metals with an arc produced between a continuous filler metal electrode and the work. Shielding is by inert gas.

ELECTRO SLAG WELDING – A welding process producing coalescence of metals with molten slag that melts the filler metal and the surfaces of the workpieces. The weld pool is shielded by slag.

EMISSIVE ELECTRODE – A filler metal electrode consisting of a core of a bare electrode or a composite electrode to which a very light coating has been applied to produce a stable arc.

EXPLOSION WELDING – A solid state welding process in which coalescence is effected by high velocity movement of the workpieces, produced by a controlled detonation.

FACE SHIELD (eye protection) – Device positioned in front of the eyes and over all or a portion of the face to protect the eyes and face.

FILLET WELD – A weld of approximately triangular cross-section joining two surfaces approximately at right-angles to each other in a lap joint, T joint or corner joint.

FILLER METAL – Metal to be added in making a welded, brazed or soldered joint.

FIRE CRACKER WELDING – Shielded metal arc welding process in which a length of covered electrode is placed along the joint in contact with the workpieces. During welding, a stationary electrode is consumed as the arc travels the length of the electrode.

FISH EYE – A discontinuity found on the fracture surface of a weld in steel. It consists of a small pore or inclusion surrounded by an approximately round bright area.

FLOW WELDING – Process which produces coalescence of metals by heating them with molten filler metal poured over the surfaces to be welded until the welding temperature is attained and until sufficient filler metal has been added.

FLUX – Material used in welding to prevent, dissolve or facilitate removal of oxides and other undesirable surface substances.

FLUX CORED ARC WELDING – Process in which coalescence of metals is effected by heating them with an arc between a continuous filler metal (consumable) electrode and the work. Shielding is by the flux contained within the tubular electrode.

FLUX CORED ELECTRODE – A composite hollow filler metal electrode containing within it ingredients to provide such functions as shielding atmosphere, deoxidation, arc stabilization and slag formation.

FOREHAND WELDING – A welding technique in which the welding torch or gun is directed toward the progress of welding.

FORGE WELDING – Process that produces coalescence of metals by heating them in air in a forge and by applying pressure or blows sufficient to cause permanent deformation at the surface.

FUSION – Melting together of filler metal and base metal (substrate) or of base metal only which results in coalescence.

FUSION WELDING – Arc welding process that uses fusion of the base metal to make the weld.

FUSION ZONE – Area of base metal melted as determined on the cross-section of a weld.

GAS METAL ARC CUTTING – Process in which metals are severed by melting them with the heat of an arc produced between a continuous filler metal electrode and the workpiece. Shielding is obtained entirely from an externally supplied gas.

GAS CARBON ARC WELDING – Process which produces coalescence of metals by heating them with an electric arc between a carbon electrode and the workpiece. Shielding is effected by a gas or gas mixture.

GAS METAL ARC WELDING – Process that produces coalescence of metals by heating them with an arc between a continuous filler metal electrode and the workpiece. Shielding is obtained by external gas supply.

GAS REGULATOR – A device for controlling the delivery of gas at some substantially constant pressure.

GAS TUNGSTEN ARC CUTTING – Process in which materials are severed by melting them with an arc produced by a tungsten electrode and the workpiece. Shielding is effected by a gas.

GAS TUNGSTEN ARC WELDING – Process that produces coalescence of metals by heating them with an arc produced between a tungsten (non-consumable) electrode and the workpiece. Shielding is obtained from a gas.

GAS WELDING – Process in which heat is furnished by a flame resulting from the combustion of a fuel gas, such as acetylene or hydrogen with oxygen; oxyacetylene being capable of producing the highest temperature flame is the most used.

GLOBULAR TRANSFER – Transfer of molten metal in large drops from a consumable electrode across the arc in arc welding.

GOUGING – The forming of a bevel or groove by material removal.

HAND SHIELD – A protective device, used in arc welding, for shielding the eyes, face and neck. A hand shield is equipped with a suitable filter plate and is designed to be held by hand.

HEAT AFFECTED ZONE – The portion of the base metal that has not been melted, but whose mechanical properties or microstructure have been altered by the heat of welding, brazing, soldering or cutting.

HELMET (eye protection) – Device designed to be worn on the head to protect eyes, face and neck from arc radiation, radiated heat, spatter, or other harmful matter expelled during arc welding.

HOT PRESSURE WELDING – Process that produces coalescence of metals with heat and application of pressure sufficient to produce macrodeformation of the base metal. Vacuum or other shielding media is used.

HOTWIRE WELDING – Arc welding process in which a filler metal wire is resistance heated by current flowing through the wire as it is fed into the weld pool.

INERT GAS – A gas which does not normally combine chemically with the base metal or filler metal.

INTERMITTENT WELD – A weld in which the continuity is broken by recurring unwelded spaces.

JOINT – Junction of members or the edges of members which are to be joined or have been joined.

JOINT EFFICIENCY – The ratio of the strength of a joint to the strength of the base metal expressed in per cent.

KERF – Width of the cut produced during a cutting process.

LAP JOINT – A joint formed between two overlapping members in parallel planes.

LASER BEAM CUTTING – Process that severs materials by melting or vaporizing them with the heat obtained from a laser beam, with or without the application of gas jets to augment the removal of material.

LASER BEAM WELDING – Process that produces coalescence of materials with the heat obtained from the application of a concentrated coherent light beam impinging upon the joint.

LIGHTLY COATED ELECTRODE – A filler metal electrode consisting of a metal wire with light coating applied subsequent to the drawing operation, primarily for stabilizing the arc.

MACHINE OXYGEN CUTTING – Cutting with an equipment that performs the operation under the constant observation and control of an oxygen cutting operator.

MACHINE WELDING – Welding with an equipment which performs the operation under the constant observation and control of a welding operator.

MANUAL WELDING – Welding performed and controlled completely by hand.

MELTING POINT – The temperature at which a metal melts.

MELTING RANGE – Temperature range between solidus and liquidus.

MELTING RATE – The weight or length of electrode melted in unit time.

METAL ARC CUTTING – Processes that sever metals by melting them with the heat of an arc between a metal electrode and the base metal.

METAL BATH BRAZING – is a dip process wherein the filler metal is obtained from the molten metal bath. This is confined to joining comparatively small work such as joints in wire.

METAL CORED ELECTRODE – A composite filler metal electrode consisting of a metal tube or other hollow configuration containing alloying ingredients.

METAL ELECTRODE – A filler or non-filler metal wire or rod, either bare or covered, used in an arc welding or cutting.

METAL POWDER CUTTING – An oxygen cutting process that severs metals through the use of powder such as iron, to facilitate cutting.

NEUTRAL FLAME – An oxyfuel gas flame in which the proportion used is neither oxidizing nor reducing.

NON-DESTRUCTIVE TESTING – Testing of welds or metal without causing any damage to the item being tested.

OVER WELDING – Depositing more filler metal than required.

OXYACETYLENE CUTTING – Process that severs metals by the chemical reaction of oxygen with the base metal at elevated temperatures caused by combustion of acetylene with oxygen.

OXYACETYLENE WELDING – Process that produces coalescence of metals by heating them with the flame obtained by combustion of acetylene with oxygen.

OXY FUEL CUTTING – A group of cutting processes used to sever metals by means of the chemical reaction of oxygen with the base metal at elevated temperatures produced by flames obtained from combustion of fuel gas and oxygen *e.g.*, oxy hydrogen cutting, oxy natural gas cutting, oxy propane cutting.

OXY FUEL GAS WELDING – A group of welding processes that produces coalescence by heating materials with an oxyfuel gas flame or flames with or without the application of pressure and with or without the use of filler metal *e.g.*, oxy hydrogen welding.

OXYGEN LANCE CUTTING – An oxygen cutting process used to sever metals with oxygen supplied through a consumable lance. The preheat to start cutting is obtained by other means.

PLASMA ARC CUTTING – Process that severs metal by melting a localized area with a constricted arc and removing the molten material with high velocity jet of hot, ionized gas issuing from the constricted orifice.

PLASMA ARC WELDING – Process that produces coalescence of metals by heating them with a constricted arc between an electrode and the workpiece (transferred arc) or the electrode and the constricting nozzle (non-transferred arc). Shielding is obtained from the hot ionized gas issuing from the orifice which may be supplemented by an auxiliary source of shielding gas.

PLUG WELD – A weld made in a circular hole or one member of a joint, fusing that member to another member.

POROSITY – Cavity type discontinuities formed by gas entrapment during solidification.

POST HEATING – Application of heat to an assembly after a welding, brazing, soldering, thermal spraying or thermal cutting.

PRE HEATING – Application of heat to the base metal immediately before welding, brazing, soldering, thermal spraying and cutting.

PRESSURE GAS WELDING – An oxyfuel gas welding which produces coalescence simultaneously over the entire area of faying surfaces by heating them with gas flames obtained from combustion of a fuel gas with oxygen and by the application of pressure, without the use of filler metal.

PROJECTION WELDING – is a modification of spot welding in which the current and pressure are localized at the weld section by the use of embossed, machined, or coined projections on one or both pieces of the work.

PROTECTIVE ATMOSPHERE – Gas or vacuum envelope surrounding the workpieces used to prevent or facilitate removal of oxides and detriment surface substances.

RANDOM INTERMITTENT WELDS – Intermittent welds on one or both sides of a joint in which the weld increments are made without regard to spacing.

PULSED POWER WELDING – An arc welding process variation in which the power is cyclically programmed to pulse so that effective but short duration values of a parameter can be utilized. Also called PULSED VOLTAGE or PULSED CURRENT WELDING *e.g.*, gas metal arc welding (pulsed arc), gas tungsten arc welding (pulsed arc).

REDUCING ATMOSPHERE – A chemically active protective atmosphere which at elevated temperature will reduce metal oxides to their metallic state.

REDUCING FLAME – A gas flame having a reducing effect owing to excess fuel gas.

RESIDUAL STRESS – Stress present in a member that is free to external forces or thermal gradients.

RESISTANCE BRAZING – is an electric brazing process wherein the heat is obtained by passing an electric current through the parts being brazed.

RESISTANCE WELDING – In this, the metal parts to be joined are heated by their resistance to the flow of an electric current *e.g.*, spot welding, seam welding.

ROLL WELDING – A solid state welding process that produces coalescence of metals by heating and by applying pressure with rolls sufficient to cause deformation at the faying surfaces.

SEAL WELD – Any weld designed primarily to provide a specific degree of tightness against leakage.

SEAM WELD – A continuous weld between or upon overlapping members, in which coalescence may start and occur on the faying surfaces, or may have proceeded from the outer surface of one member.

SEMI AUTOMATIC ARC WELDING – Arc welding with equipment that controls only the filler metal feed. The advance of the welding is manually controlled.

SEMI BLIND JOINT – A joint in which one extremity of the joint is not visible.

SERIES SUBMERGED ARC WELDING – A submerged arc welding process variation in which electric current is established between two consumable electrodes which meet just above the surface of the workpieces which are not part of the electric circuit.

SHIELDED CARBON ARC WELDING – A carbon arc welding process in which shielding is effected by the combustion of a solid material fed into the arc or from a blanket of flux on the workpieces or both.

SHIELDED METAL ARC CUTTING/WELDING – Cutting or welding effected by heating the workpieces with an arc between a covered metal electrode and the workpiece. Shielding is obtained from decomposition of the electrode covering,

SHIELDING GAS – Protective gas used to prevent atmospheric contamination.

SHORT CIRCUITING TRANSFER ARC WELDING – Metal transfer in which molten metal from a consumable electrode is deposited during repeated short circuits.

SHRINKAGE VOID – A cavity type discontinuity normally formed by shrinkage during solidification.

SILVER BRAZING – Brazing process that employs silver alloys. Also called hard soldering or silver soldering.

SLAG INCLUSION – Non-metallic material entrapped in weld metal or between weld metal and base metal.

SLOT WELD – A weld made in an elongated hole in one member of a joint fusing that member to another member. The hole may be open at one end.

SOLDER – A filler metal used in soldering which has a liquidus not exceeding 450°C.

SOLDERING – A group of welding processes that produces coalescence of materials by heating them to a suitable temperature and by using a filler metal having a liquidus not exceeding 450°C and below the solidus of the base metals.

SOLID STATE WELDING – A group of welding processes that produces coalescence at temperatures essentially below the melting point of the base metal being joined, without the addition of a brazing filler metal. Pressure mayor may not be used.

SPATTER – The metal particles expelled during fusion welding, that do not form a part of the weld.

SPELTER – is common brass, the first material used to make a brazed joint in ferrous material.

SPOT WELD – Weld made between or upon overlapping members in which coalescence may start and occur on the faying surfaces or may proceed from the surface of one member. The weld cross section (plan) is approximately circular.

SPRAY TRANSFER ARC WELDING – Metal transfer in which molten metal from a consumable electrode is propelled axially across the arc in small droplets.

STACK CUTTING – Thermal cutting of stacked metal plates arranged so that all the plates are severed by a single cut.

STAGGERED INTERMITTENT WELD – An intermittent weld on both sides of a joint in which the weld increments on one side are alternated with respect to those on the other side.

STRINGER BEAD – A type of weld bead made without appreciable weaving motion.

STUD WELDING – A general form for the joining of a metal stud or similar part to a workpiece. Welding may be effected by arc, resistance, friction or other suitable process, with or without external gas shielding.

SUBMERGED ARC WELDING – An arc welding process in which the arc and molten metal are shielded by a blanket of granular, fusible material on the workpieces.

SURFACING – Application by welding, brazing or thermal spraying of a layer(s) of material to a surface to obtain desired properties or dimensions as opposed to making a joint.

TACK WELD – A weld made to hold parts of a weldment in proper alignment until final welds are made.

THERMIT MIXTURE – A mixture of metal oxide and finely divided aluminium with the addition of alloying metals as required.

THERMIT WELDING – Welding process that produces coalescence of metals by heating them with superheated liquid metal from a chemical reaction between a metal oxide and aluminium, with or without the application of pressure.

T-JOINT – A joint between two members located approximately at right-angles to each other in the form of a T.

TORCH BRAZING – A brazing process in which the heat required is furnished by a fuel gas flame.

TWIN CARBON ARC WELDING – Process that produces coalescence of metals by heating them with an electric arc between two carbon electrodes. No shielding is used.

UNDER WELDING – Depositing less filler material than required.

WEAVE BEAD – A type of weld bead made with transverse oscillation.

WELD – A localized coalescence of metals or non-metals produced either by heating the materials to the welding temperature, with or without the application of pressure or by the application of pressure alone and with or without the use of filler material.

WELDING ELECTRODE – A component of the welding circuit through which current is conducted and which terminates at the arc, molten conductive slag or base metal.

CHAPTER 11

HEAT AND SURFACE TREATMENTS

AGING – The process of holding metals at room temperature or at a predetermined temperature for the purpose of increasing their hardness or strength by precipitation. Aging is also used to increase dimensional stability in metals such as castings.

AGE HARDENING – Change in the physical properties, *e.g.*, hardness and strength, that occurs in certain light metals after heat treatment.

ALLOYING OF METAL – The addition of varying proportions of other elements to a basic metal in order to produce an alloy having specific properties.

ANODIC TREATMENT – Formation of a protective layer of oxide on the surface of aluminium and its alloys by electrolytic action, in order to resist corrosion.

ANNEALING – A heat treatment in which metals are heated and then cooled very slowly for the purpose of decreasing hardness. Annealing is used to improve machinability and to remove stresses from weldments, forgings and castings. Also used to remove stresses resulting from cold working and to refine and make uniform the microscopic internal structures of metals.

ANODIZING – To subject a metal to electrolytic action, as takes place at the anode of a cell, in order to coat it with a protective or decorative film, used for nonferrous metals.

AUSTEMPERING – A heat treating process consisting of quenching a ferrous alloy at a temperature above the transformation range in a medium such as molten lead, the temperature of the quenching medium is maintained below that of pearlite and above that of martensite formation to produce a tough, hard microstructure.

AUSTENITIZING – The process of forming austenite (a solid solution of iron and carbon and sometimes other elements).

BLACK OXIDE COATING – Coating produced by converting the surface of iron or steel to black iron oxide having a thickness of about 0.0025 mm.

CADMIUM PLATING – Electroplating process for the application of cadmium to steel and iron parts to prevent rust.

CALORIZING – Rust proofing process for ferrous metals in which an aluminium film is formed on the surface of the metal. Means of protecting iron from oxidation at elevated temperatures.

CARBURIZING – A process that introduces carbon into a heated solid ferrous alloy by having it in contact with a carbonaceous material. The metal is held at a temperature above the transformation range for a period of time. This is generally followed by quenching to produce a hardened case.

CASE HARDENING – Heating a steel in the presence of a solid, liquid or gas, rich in carbon, in order to enable the surface to be hardened, while retaining a tough, ductile core.

CHROMIUM PLATING – Electrolytic deposition of chromium on a metal surface, as a protection against corrosion, to provide improved wearing properties, or to build up an undersize part.

CHROMIZING – Similar to carburizing. Low carbon steel parts are packed with a mixture of alumina and chromium powder and heated in a hydrogen atmosphere, forming a surface layer of chromized material of 10 to 20% chromium, according to time and temperature of heating.

COLOURING METAL – Production of a coloured surface on a metal or alloy by the use of heat or chemical action, or by a combination of the two.

COSLETTIZING – Rust proofing process in which an iron phosphate skin is formed on the surfaces of ferrous parts, the skin follows even the microscopic irregularities, yet is remarkably tough and corrosion resistant.

DECARBONIZATION – The loss of carbon from the surface of a ferrous alloy as a result of heating it in the presence of a medium such as oxygen that reacts with the carbon.

ELECTROLYTIC POLISHING – Method of polishing metals in which the work forms the anode of an electrical circuit, and is suspended in a suitable bath of acid.

ELECTROPLATING – Deposition of a metal on a surface by electrolytic action.

FLAME HARDENING – Process of hardening by which steel or cast iron is raised to a high temperature by a gas torch flame and then almost immediately quenched.

GALVANIZING – Rust prevention treatment which consists of coating the metal (iron or steel) with a fairly thick film of zinc.

HARDENING – Process of increasing the hardness of a ferrous alloy by austenitizing and quenching, also the process of increasing the hardness of some stainless steels and non-ferrous alloys by solution heat treatment and precipitation.

HARD SURFACING – Arc welding a surface layer usually at strategic spots on a part, using an electrode having a suitable alloy content for the purpose. Also known as SURFACING and FACING.

HEAT TREATMENT – Operation or combination of operations, involving the heating and cooling of a metal or alloy in its solid state with the object of changing the characteristics of the material.

HOT ROLLING – Surface layers of noble metals are often hot rolled (or inlayed) on less noble metals.

INDUCTION HARDENING – Heating the surface of cast iron or tool steel by means of electromagnetic currents followed by a quench.

LACQUERING – A protective coat given to an article to prevent the polished surface from tarnishing, to prevent oxidation or to improve the general appearance and make the article more pleasing to the eye, and hence more saleable.

MARTEMPERING – The process of quenching an austenitized ferrous alloy to a temperature just above or near the M's point and maintaining until the temperature throughout the part is uniform. The alloy is then allowed to cool slowly in air through the range of martensitic formation.

METALLIZING – Metal spraying process in which the coating metal is melted (either in a gas flame or in a separate melting pot), then atomized by an air blast and deposited in atomized form on the surface to be coated.

NEGRADIZING – Rust proofing process in which a ferrosoferric oxide surface is produced on steel or iron.

NICKEL PLATING – Deposition of nickel on a metal by electrolytic action to provide a protective surface or to build up the surface of a worn or undersize part.

NITRIDING – A process of case hardening in which a special ferrous alloy is heated in an atmosphere of ammonia or is in contact with any other nitrogenous material. By this, surface hardening is achieved by the absorption of nitrogen without quenching.

NORMALIZING – Process applied to iron base alloys, such as steel, to refine the grain structure and remove the effects of previous processing for example hot rolling.

ORGANIC FINISHING – Coating a surface with a continuous film of an organic material for protecting the surface from corrosive influences, for enhancing the appearance or a combination of both.

PARKERIZING – A coslettizing process in which a patented mixture of iron and manganese phosphates is used in solution to provide a rust proof surface on iron and steel parts.

PHOSPHATE COATING – Coating produced by converting the surface of iron or steel parts to insoluble phosphate, such as iron phosphate or zinc iron phosphate.

PICKLING – Treatment of objects with an acid solution to remove all oxide, scale or dirt. Usually done to clean and brighten the surface, although sometimes used for etching.

POWDER COATING – Surface coating of almost any metal or alloy applied to another by powder method.

PRECIPITATION HARDENING – A process of hardening an alloy by heat treatment in which a constituent precipitates from a supersaturated solid solution while at room temperature or at some slightly elevated temperature.

QUENCHING – Rapid cooling of a metal in a bath of fluid during heat treatment after it has been heated to a given temperature, thus trapping the molecules of the metal in the desired structure.

RECRYSTALLIZATION – A process in which the distorted grain structure of metals that are subjected to mechanical deformation is replaced by a new strain free grain structure during annealing.

RUST PROOFING – Treatment of iron and steel parts to render them resistant to rust, implies a more permanent form of protection.

SHERADIZING – Heat treatment process in which zinc dust is used for producing a rust proof surface on ferrous metals. It derives its title from the name of the inventor SHERARD OSBORN COWPER COLES.

SILICONIZING – Process in which iron and steel parts are impregnated with silicon for improving the surface resistance to corrosion, heat and wear.

SOAKING – A prolonged heating of a metal at a predetermined temperature to create a uniform temperature throughout its mass.

SOLUTION HEAT TREATMENT – A process in which an alloy is heated to a predetermined temperature far a length of time that is suitable to allow a certain constituent to enter into solid solution. The alloy is then cooled quickly to hold the constituent in solution, causing the metal to be in an unstable supersaturated condition. This condition is often followed by age hardening.

STRAIN HARDENING – An increase in hardness and strength of a metal that has been deformed by cold working or at temperatures lower than the recrystallization range.

SUBZERO HEAT TREATMENT – Method of heat treatment of steel, introducing a cooling period or periods at a temperature well below freezing point, the object being to ensure complete transformation of austenite to martensite.

TEMPERING STEEL – Heat treatment designed to relieve the stresses and brittleness set up in carbon or alloy steel tools and other parts after hardening and to restore the required degree of toughness and ductility to hardened steel.

TINNING – Application of a layer of tin to the surface of another metal, either as a protection against corrosion or as a preliminary to soldering.

WORK HARDENING – Phenomena in which the grains become distorted and elongated in the direction of working (rolling). This process, also called STRAIN HARDENING, hardens and strengthens metals but reduces their ductility.

ZINC PLATING – Electroplating process for the deposition of zinc on metallic surfaces.

CHAPTER 12

MECHANICS OF MACHINES

ABSOLUTE MOTION – Motion of a body in relation to some other body which is at rest.

ACCELERATION – Rate of change of velocity with respect to time, of a particle which is in motion. It is a vector quantity.

ADDENDUM – The radial distance from the pitch circle to the top of the tooth.

ANGULAR ACCELERATION – The time rate of change of angular velocity.

ANGULAR VELOCITY – The time rate of change of angular displacement of a point rotating about a fixed axis (expressed in radians per unit time) Angular velocity of a machine part is often expressed in revolutions per minute (RPM) and is denoted by n.

ANGULAR VELOCITY OF PRECISION – The rate of change in the direction of the plane of rotation of a rotating disc.

ARC OF CONTACT – The arc traced out along the pitch circle while one pair of teeth of gear wheels is in contact (divided into arc of approach and arc of recess).

BEVEL GEARING – Gearing arrangement in which the axes of the shafts connected by gears intersect.

CAM – A reciprocating, oscillating or rotating body which imparts reciprocating or oscillating motion to a second body, called the FOLLOWER with which it is in contact.

CAM PROFILE – The surface profile of the cam that decides the desired motion of the follower.

CENTRIFUGAL FORCE – Radial outward force acting on a body moving along a circular path with uniform velocity.

CENTRIFUGAL GOVERNOR – The effort of the governor is obtained from the change in centrifugal force on (usually) two rotating masses, known as balls, when an increase or decrease in the governor speed occurs.

CENTRIPETAL FORCE – The force that must act radially inward in order to constrain a particle to follow a curved path at uniform velocity.

CIRCULAR PITCH – Length of arc round the pitch circle between the corresponding points on adjacent teeth of a gear.

COMPLEX MECHANISMS – Mechanisms which have two or more floating links.

COMPOSITION OF VECTORS – Composition refers to the adding together of any number of vectors. The sum is called their resultant and the vectors are called the components of the resultant.

COMPOUND CHAIN – A kinematic chain in which there are more than four pairs.

COMPOUND GEAR TRAIN – A gear train containing compound gears *i.e.*, gears, two or more in number integral with one another being used on the same shaft.

COMPOUND PENDULUM – A rigid body suspended vertically so as to oscillate with small amplitude under the action of gravity.

CONSERVATION OF ENERGY – The total energy possessed by a system of moving bodies is at every instant constant, provided no energy is rejected to or received from a source external to the system.

CONSERVATION OF MOMENTUM – For a system of moving bodies which is not acted upon by any external forces, the sum of the moments remain constant.

CONTROLLING FORCE OF A GOVERNOR – The inward radial force exerted on each ball of a centrifugal governor by the arms, springs etc., which are attached to it.

CURVILINEAR MOTION – A translation in which points in the body move along curved path (motion of a wheel).

CYCLE OF MOTION – Motion of a mechanism when it moves through all its possible configurations and returns to its starting position. The time required for one cycle is called PERIOD.

CYCLOIDAL TEETH – Profile of the teeth formed by the locus of a point on a circle rolling on the inside (for the flank) and on the outside (for the face) of the pitch circle.

CYLINDRICAL CAM – Type of cam in which the motion of the follower is controlled by a path traced out on the surface of a cylinder which is rotating about its axis.

DEAD WEIGHT GOVERNOR – The governor in which the radius of the ball path is controlled by levers and weights, the latter being usually attached to the control sleeve.

DEDENDUM – The radial distance from the pitch circle to the bottom of the tooth space.

DIAMETRAL PITCH – Number of teeth per inch diameter.

DISC CAM – An irregular disc rotating about a fixed axis and imparting reciprocating or oscillating motion to a follower in a plane at right angles to the cam axis.

DYNAMICS OF MACHINES – Treatment with the forces acting on the parts of a machine and the motions resulting from these forces.

DYNAMOMETER – A device for measuring the forces or couples which tend to change the state of rest or of uniform motion of a body.

ELLIPTIC TRAMMEL – An instrument used for drawing ellipses.

ENERGY – Capacity for doing work.

EPICYCLIC GEAR TRAINS – Gear trains in which the axis of one or more gears moves relative to the frame. The gear at the centre is called the SUN, and gears whose axes move are called PLANETS. Also called PLANETARY GEARS.

EPICYCLOID – The locus of a point on the circumference of a circle which rolls outside a circular arc, without slipping.

FLOATING LINK – A link in a mechanism which does not have a fixed center of rotation (*e.g.*, coupler in a four bar linkage).

FORCE – The entity which when acts on a body can cause a change in its velocity or direction or both.

FRAME – That part of a machine which is stationary and which supports the moving parts.

FRICTION DRIVE – Drive in which the rotation of one body causes another body in contact with it to rotate due to sufficient friction between the bodies.

GEAR CLEARANCE – The radial distance from the top of the tooth to the bottom of the tooth space in a mating gear unit.

GEAR TRAIN – Unit composed of two or more gears in mesh for the purpose of transmitting motion from one shaft to another.

GOVERNOR – Device that controls the mean speed of an engine over a period of time, as distinct from the flywheel, which limits the fluctuation of speed during one cycle but is not able to prevent a change in mean speed from cycle to cycle.

GOVERNOR EFFORT – Mean force exerted at a sleeve due to a 1% change in speed of governor.

GOVERNOR POWER – Work done at the sleeve for a 1% change in speed, equal to the governor effort times the sleeve displacement.

GYROSCOPIC ACCELERATION – The rate of change of angular velocity of precision of a rotating disc.

HELICAL MOTION – Motion of a body in which each point in the body describes a helix. Helix is the locus of a point which rotates about an axis at a fixed distance and at the same time moves parallel to the axis.

HELICAL GEARING – A type of spur gearing in which although the axes of the shafts are parallel, the teeth are cut on helices instead of straight across the wheels parallel to the axis.

HIGHER PAIRS – Types of kinematic pairs, namely, two elements generally have line or point contact and the pair must be force closed in order to provide completely constrained motion.

HUNTING OF GOVERNOR – The governor is said to hunt if the engine speed is caused to fluctuate continually above and below the mean speed.

HYPOCYCLOID – The locus of a point on the circumference of a circle which rolls inside a circular arc without slipping.

IMPULSE – Time integral of the impulsive force acting on a body.

IMPULSIVE FORCE – Force that acts on a body for an extremely short interval of time and makes the body to move. Occur in collisions, in explosions, in the striking of a nail by a hammer or of a pile by a tup or monkey.

INERTIA GOVERNOR – Governor in which the position of the flyballs are affected by the rate of change of speed of the governor shaft.

INSTANT CENTRE – (1) A point in one body about which another body is rotating either permanently or at the instant (2) A point common to two bodies having the same linear velocity in both magnitude and direction in each.

INTERMITTENT MOTION MECHANISM – A linkage which converts continuous motion into intermittent motion (*e.g.*, indexing mechanism).

INVOLUTE – The locus of a point on a straight line which rolls, without slipping, on the circumference of a circle, or alternatively the locus of a point on the chord which is held taught and unwound from a cylinder.

INVOLUTE TEETH – The outline of a tooth traced out by a point on a chord unwrapped from a circle (known as base circle).

ISOCHRONISM OF GOVERNOR – A governor is said to be isochronous, if , neglecting friction, the equilibrium speed is the same for all radii of the flyballs.

KENNEDY'S THEOREM – Any three bodies having plane motion relative to one another have three instant centers, and they lie in a straight line.

KINEMATIC CHAIN – A group of links either joined together or arranged in a manner that permits them to move relative to one another.

KINEMATIC DIAGRAM – A scale drawing representing the machine so that only the dimensions which affect its motions are recorded.

KINEMATIC PAIR – Two bodies in contact, between which there is relative motion and this motion is completely constrained *e.g.*, turning pair, sliding pair, screw pair.

KINEMATICS OF MACHINES – A study of the relative motion of machine parts *e.g.*, displacement, velocity and acceleration.

KINETICS – Study which deals with the inertia force arising from the combined effect of the mass and the motion of the parts.

LINEAR ACCELERATION – The time rate of change of liner velocity.

LINEAR VELOCITY – The time rate of change of linear displacement of a point or body.

LINK – Name given to any body which has relative motion to another. Also called ELEMENT. A RIGID LINK is one whose deformations are so small that they can be neglected in determining the motions of various other links in a machine. A belt or chain is a FLEXIBLE LINK.

LOWER PAIRS – Types of kinematic pairs, namely two elements have surface contact and when relative motion takes place, the surface of one element slides over the surface of the other element.

MACHINE – A combination of resisting bodies, with successfully constrained relative motions, which is used for transmitting or transporting available energy so as to do some particular kind of work *e.g.*, electrical motor, internal combustion engine.

MASS OF A BODY – The property of a body which determines its resistance to change its velocity.

MECHANISM – A constrained kinematic chain which means the motion of anyone link will give a definite, predictable motion to each of the others.

MODULE – Reciprocal of diametral pitch.

MOMENTUM – The product of the mass and velocity of a body.

NORMAL ACCELERATION – The time rate of change of velocity of a point in a direction normal to its path. This results from a change in the direction of its linear velocity.

OLDHAM COUPLING – A mechanism for connecting two shafts having parallel misalignment. The coupling transmits a constant velocity ratio.

PAIR – Two bodies in contact constitute a pair. LOWER PAIRING exists when two surfaces are in contact. HIGHER PAIRING refers to the contact which exists at a point or along a line.

PANTOGRAPH – Mechanism used to reproduce to an enlarged or reduced scale and as exactly as possible the path described by a given point.

PARALLEL MECHANISMS – Linkages which give parallel motion (*e.g.*, pantograph which is used for reducing or enlarging drawings and maps, also used for grinding cutting tools or cutting torches to duplicate complicated shapes).

PATH OF CONTACT – The path traced out by the point of contact between a pair of teeth (may be divided into approach and recess).

PINION – The small of the two mating gear wheels.

PITCH CIRCLES – Equivalent rolling circles for a pair of mating gears.

PITCH CIRCLE DIAMETER – The diameter of a circle which by a pure rolling action would transmit the same motion as the actual gearwheel.

PITCH LINE – The point of contact of two circles.

PITCH SURFACES – The cylindrical surfaces of the equivalent rolling circles for a pair of mating gears.

PLANE MOTION – A body has plane motion if all the points move in planes which are parallel to some reference plane (called plane of motion).

POSITIVE DRIVE – The drive that exists in a direct contact mechanism if motion of the driving link compels the follower to move (*e.g.*, cam and follower).

POWER – Rate of doing work or work done in unit time.

PRECESSIONAL MOTION – The change in the direction of the plane of rotation of a rotating disc.

PRESSURE ANGLE – Angle between the common normal and the tangent at the pitch point in a gear drive. Also called ANGLE OF OBLIQUITY.

QUICK RETURN MECHANISM – The mechanism used in machine tools such as shapers and power driven saws for the purpose of giving the reciprocating cutting tool a slow cutting stroke and a quick return stroke with a constant velocity of the driving crank.

RACK – A portion of a gear wheel which has an infinitely large number of teeth.

RACHETS – Mechanisms used to transform motion of rotation or translation into intermittent rotation or translation.

RECTILINEAR MOTION – A motion wherein all points of the body move in straight line paths (*e.g.*, piston motion).

RELATIVE INSTANTANEOUS CENTER – In the case of two bodies, it is the point about which either of them appears to turn (at that instant) if the other is considered fixed (*e.g.*, if two links in a mechanism are pinned together, the pin becomes the relative instantaneous center, if the two bodies are in pure rolling contact, the point of contact is the relative instantaneous center).

RELATIVE MOTION – A body has motion relative to another body only if there is a difference in their absolute motions.

RESOLUTION OF VECTORS – Resolution refers to the breaking down of a vector into any number of component vectors.

REVERTED GEAR TRAIN – The compound gear train in which the first and the last gears are coaxial (*e.g.*, units used in automobile transmission, lathe back gears, industrial speed reducers and in clocks).

ROLLING CONTACT – In a direct contact mechanism, rolling contact exists only if there is no sliding and hence the tangential components of velocities of the contact point on the two bodies are equal in magnitude and direction.

ROTATION – In rotation all points in a body remain at fixed distances from a line which is perpendicular to the plane of motion. This line is the AXIS OF ROTATION.

SCALAR QUANTITIES – Those quantities which have magnitude only (and no direction) *e.g.*, distance, area, volume and time.

SENSITIVITY OF GOVERNOR – Ratio of the mean speed to the speed range of the governor over its limits of operation.

SIMPLE AND COMPOUND MECHANISM – A simple mechanism consists of three or four links. All other mechanisms, or those consisting of more than four links are compound mechanisms. Compound mechanisms are usually made up of combinations of simple mechanisms.

SIMPLE GEAR TRAIN – A gear train in which there is only one gear on each shaft.

SIMPLE HARMONIC MOTION – A particle having rectilinear motion has simple harmonic motion if its acceleration is proportional to the displacement of the particle from a fixed point and is of opposite sign.

SKEW GEARING – Gearing arrangement in which the axes of the shafts connected by gears are non parallel and non intersecting.

SLIDING CONTACT – Sliding exists in a direct contact mechanism whenever the bodies have relative motion along the tangent through their point of contact.

SPEED – The rate of change of magnitude of displacement with respect to time.

SPHERICAL MOTION – A point has spherical motion if it moves in three dimensional space and remains at a fixed distance from some fixed point. A body has spherical motion if each point in the body has spherical motion.

SPIRAL GEARING – A type of skew gearing, but differs in one respect *i.e.*, the contact between pitch surfaces is point contact instead of line contact.

SPRING LOADED GOVERNOR – The governor in which the control of the flyballs is by springs operated directly on the balls or on the sleeve.

SPUR GEARING – Gearing arrangement in which the axes of the shafts connected by gears are parallel and the teeth are cut parallel to the axes.

STABILITY OF GOVERNOR – The governor is said to be stable if there is one equilibrium speed for each radius of rotation of the flyballs and this speed increases with the radius.

STATICS – Study which deals with forces which act on the various parts, when these parts are assumed to be without mass.

STRAIGHT LINE MECHANISMS – Linkages having a point that moves along a straight line or nearly along a straight line, without being guided by a plane surface (*e.g.*, Watts mechanism, Scott Russell mechanism).

TANGENTIAL ACCELLERATION – The time rate of change of velocity of a point in a direction tangent to its path. This results from a change in its linear velocity.

TRANSLATION – A body has translation if it moves so that all straight lines in the body move in parallel positions.

UNCONSTRAINED KINEMATIC CHAIN – An arrangement of links wherein for a given motion of one of the links, each of the others is not constrained to move in a definite predictable manner.

UNIVERSAL JOINT – Joint used to connect intersecting shafts (*e.g.*, Hooke or cardan joint).

VECTOR QUANTITIES – Those entities which have magnitude and direction. (*e.g.*, displacement, velocity, acceleration and force). Vector quantities are represented by a straight line with an arrow head (magnitude is represented by its length and direction by the arrow head).

VELOCITY – When a particle is in motion, the rate of change of its displacement with respect to time. It is a vector quantity.

VELOCITY RATIO OF GEAR TRAIN – Ratio of the angular velocity of the first gear in the train to the angular velocity of the last gear.

WORM GEARING – A form of special gearing in which the axes of the driving and driven shafts are usually at right angles and the velocity ratio is high, the driving gear being of smaller diameter.

WORK – Product of the force acting on a body and the displacement of the body caused by that force.

WORKING SURFACE – With respect to a gear, the working surface above the pitch surface is called the FACE OF THE TOOTH and that below the pitch surface is called the FLANK OF THE TOOTH.

CHAPTER 13

MATERIAL HANDLING

ADJUSTABLE RAMP – A loading platform that is power operated or mechanically operated.

APRON CONVEYOR – A conveyor for transporting packages or bulk materials, consisting of a series of metal or wood slats (also rubber, cotton, felt wire etc.) attached to an endless chain. Also called SLAT CONVEYOR.

ARRESTING GEAR – Means of sustaining the load which do not interfere with the hoisting gear but prevent it from coming down due to gravity *e.g.*, ratchet and pawl arrangements and friction type.

AUTOMATIC GRAB – A crane grab in which the grasping and releasing of the load are effected without manual assistance.

BAND or BELT CONVEYOR – An endless band passing over, and driven by horizontal pulleys, thus forming a moving track which is used to convey loose material or small articles.

BALANCED LUFFING – Luffing mechanism, in which the moment due to the weight of the jib is at balance with the moment produced by the counterweight.

BARGES AND LIGHTERS – Shallow draft, box like vessels used for cargo transport in protected waters such as bays, rivers and canals.

BARREL ELEVATOR – This comprises parallel travelling chains, with curved arms projecting. The chains pass over sproket wheels at the top and bottom of the elevator, and lift barrels from a loading platform to a runway.

BARREL HOPPER – A machine for unscrambling, orientating and feeding small components during a manufacturing process, in which a revolving barrel tumbles the components onto a sloping, vibrating feeding blade.

BAY – An area used for the open storage of heavy items.

BELT CONVEYOR – A conveyor which consists of a belt of suitable material such as rubber, canvas, balata etc., running over a pair of end drums or pulleys and supported at intervals by a series of rollers called idlers, these in turn being supported on a conveyor frame.

BELT CONVEYOR IDLERS – Number of idler rolls provided between the terminal pulleys to prevent the belt from sagging due to gravity and under the load.

BELT CONVEYOR PULLEYS – Wheels used to support and drive the belt. They include drive, terminal or bend, take up and snub pulleys.

BIN – An enclosed space for storing certain types of goods.

BIN TICKET – Tickets attached to storage places to provide information on the quantity of goods received, issued and on hand.

BRACING – Securing the contents of a shipment to prevent shifting and damage.

BRAKE – Arrangement in the hoisting machinery to stop the load and hold it when applied to the hoisting motion or bring the relevant mechanisms at rest within specified braking distances. May be a band brake, disc brake or a cone brake.

BUCKET ELEVATOR – Conveyor equipped with buckets which carry bulk material in the vertical or near vertical direction, loading is at the bottom and discharging is at the top.

BULK COMMODITY TRUCK – Trucks used to transport loose bulk materials, such as sand and gravel.

BUCKET OR SKIP HOISTS – Hoisting equipment for handling of bulk materials in self dumping buckets or skips.

BULLDOZER – A pendant attachment mounted on crawler and wheel tractors, that strips off soil surface and transports it to the required spot.

CANTILEVER RACKS – Racks supported only on one end, leaving the other end open for placing and removing the long bars, rods etc.

CAPSTAN DRUM – Equipment used for hoisting anchors on board ships, hauling various loads etc.

CAR PULLERS – Equipment used for shunting railroad cars at ports and docks.

CAROUSEL – A rotating or circulating storage device. The worker stays in one place while the needed item comes to the work station.

CHAIN CONVEYORS – Conveyors which employ chains of various designs as the driving traction element. Chain conveyors carry aprons, pans, buckets, cradles, pockets, cars etc.

CHOCKS – Supports used to keep boxes off the ground.

CHUTE – An inclined surface with sides for material movement by gravity.

COLD STORAGE WAREHOUSE – A warehouse used for storing highly perishable goods and foods.

CONTAINERIZATION – Transportation of freight in sealed portable containers.

CONVEYOR – A mechanism that moves material along a fixed path. A mechanical device for carrying packages or bulk material from place to place (as by an endless moving belt or a chain of receptacles).

COUNTER WEIGHTS – Weights provided on cranes to offset the dead weight of metal structure and, to a certain extent, the moment due to the hook load.

COVERED HOPPER – A freight car with a closed top designed to meet the needs of malt and grain shippers.

CRANE – A machine for raising, shifting and lowering heavy weights by means of projecting swinging arm or with the hoisting apparatus supported on an overhead track.

CRANE DERRICK – The distance between the loads centre of gravity and the axis about which crane boom can swing.

CRANE HELICOPTER – A crane hung from a helicopter, used as a means of moving loads and doing a variety of jobs in regions which are difficult of access, capable of descending vertically on the load and lifting it from the ground directly.

CRAWLER CRANE – A crane mounted on a crawler mounting, *e.g.*, a frame supported by track laying assemblies which obtain the drive from an engine mounted on a rotating part of the crane.

CREEPING DRIVE UNIT – An arrangement frequently employed on electric hoists to obtain extra low spotting speeds.

DEAD ZONE – In a shop floor, when two overhead bridge cranes serve adjacent bays, part of floor area (about 15 to 20%) inaccessible for any of the hooks, as close as they may be. This is dead zone.

DERRICK – A framework over a drill hole (*e.g.*, on oil well) for supporting the boring mechanism for hoisting.

DIESEL ELECTRIC DRIVE – System consisting of a diesel generator set and electric motors on all motions.

DOUBLE BOTTOM RIG – Two trailers joined to a tractor.

DOUBLE FACED PALLET – A pallet with two decks that form the top and bottom surfaces.

DOUBLE STRING TIE – A method of tying heavy packages.

DOUBLE WALL CORRUGATED FIBREBOARD – Fibreboard consisting of three flat facings and two corrugated pieces arranged in alternate layers.

DOUBLE WRAP – Two or more sheets of paper used to wrap breakable articles.

DRIVEWAY INSTALLATION – A loading ramp that raises or lowers the truck so that the floor of the truck is level with the dock.

DRUM – In hoisting unit, a device for spooling the rope in one layer or in more than one layer.

DUNNAGE – Lumber or other material used to brace a shipment in transport.

ELECTRIC DRIVE – An arrangement consisting of an electric motor, suitable motor controls, and a gear train linking the motor with the operating mechanism.

ELEVATOR – A platform or caged hoist that moves material or personnel from one level to another.

ELEVATING GRADER – Machine that cuts soil layer by layer and moves it away by means of a conveyor into a dump or onto transporting facilities.

ESCALATOR – Chain conveyor specially adapted for the vertical transportation of people over an inclined path. It is an inclined conveyor with the driving traction in the form of steeps attached to step drive chains and each escalator step is carried on four wheels riding on tracks.

EXCELSIOR – Shredded or curled pieces of wood.

FISHYBACK – The transportation of highway trailer bodies abroad ship.

FLAMMABLE – A marking on the box to indicate that the contents may easily set on fire or have a tendency to explode.

FLAT BED TRAILER – A flat surfaced platform with wheels.

FLATTENED STRAND WIRE ROPE – A modification of linear contact ropes. During its manufacture, every strand laid from wires of circular cross–section is radially squeezed over the circumference so as to flatten the wires.

FLEX – Flow tank car. A pressurized tank car that can unload in a few minutes.

FLOATING CRANE – A crane mounted on self propelled or towed pontoons, used for salvage work, along shore and offshore duties, crane motions are accomplished with electric drives, commonly of the dc type employing the Ward Leonard control.

FLOW RACK – Rack in which the material can be inserted at one side and will move by gravity to the other where order picking takes place.

FORCE OF GRAVITY – Force decided by the attraction the earth exerts on a body which, unlike the mass, depends on the local value of the acceleration of free fall.

FORK LIFT TRUCK – Equipment designed to pick up, transport, stack and unstack pallet loads.

FOURWAY PALLET – A pallet design that permits entry of materials handling equipment from four sides.

FRAGILE – A marking on the box to indicate that the contents are delicate and breakable.

GANTRY CRANE – Crane consisting of a bridge crane superimposed on two frames or legs operating on a rail track flush with the ground.

GENERAL MERCHANDISE WAREHOUSE – The most common type of public warehouse for storing manufactured or non-manufactured goods that require protection.

GONDALA – A car with low sides and no top.

GRAB – Device suspended from the crane hook which grasp and release the load (of given shape and size) at almost no time.

GRAVITY ROLLER CONVEYOR – A material handling installation in which the bed consists of rolls fitted to a frame. Suitable for handling of unit loads and containerized high volume materials.

HAND DRIVE – Manual operation which is in wide use on the hoisting and slewing motions of slow running, low capacity machinery.

HOOK – Multipurpose attachment widely used in hoisting installations, carries the load by one or more of slings made of hemp ropes or chains.

HYDRAULIC POWER EQUIPMENT – Unit which consists of an electric motor or internal combustion engine driving a pump which feeds fluid into the hydraulic operating cylinder through the lines with control valves.

H WARP – A method of wrapping large fragile items.

HOISTING MACHINE – Moves loads vertically as well as transfers them from one point of the area covered by the machinery into another.

INDUSTRIAL TRUCK – A self powered, individually operated conveyance for moving materials or persons.

JACK – Device used to lift loads through a short height some 0.8 to 1 m used widely in repair and erection jobs, *e.g.,* screw jacks, rack and lever jack's rack and pinion jacks and hydraulic jacks.

LANG LAY ROPE – A rope in which the wires are twisted in the same direction as the strands are.

LEFT HAND LAY ROPES – Ropes in which the strands bend round to the left.

LEVEL LUFFING – Luffing mechanism which ensures that the crane hook travels along an almost horizontal path when luffing is in progress.

LIFT – Hoisting equipment adapted to lift loads vertically.

LIFTING CAPACITY – The maximum safe load the machine is designed to handle.

LIFTING MAGNET – Handling device, operating on direct current. Widely used in handling steel and cast iron loads.

LIFTING TACKLE – A system of stationary and movable sheaves interlinked by a pliable member in the form of a wire rope to obtain a gain in either force (power lifting tackle) or speed (speed lifting tackle).

LUFFING or DERRICKING – It is the pivoting of the crane jib in a vertical plane so as to change the reach.

MANIPULATOR – A device that grasps an item and moves it to a new position or orientation for the next operation. It may be a simple pincer or complicated robot.

MATERIALS HANDLING – Procedures involving receiving, storing, sorting, packing, shipping, freight handling and controlling of goods and materials.

MONORAILS – An overhead track (usually an I beam) upon which the carriers or hoists move.

MULTIPLE RACK CAR – Special equipment designed to carry motor cars.

NON-SPINNING ROPES – Multilayer constructions having alternately opposite lays of individual layers.

OPEN HOPPER – A freight car with an open top and collapsible bottom.

ORDINARY LAY ROPE – A rope in which the direction of twist of wires is opposite to that of the strands in the rope.

ORDINARY (untwisting) ROPES – Ropes in which the wires and strands fail to retain their position once the wire sizings are removed from the ends and consequently, these ropes tend to untwist.

ORIGINATING CARRIER – The first carrier to receive the goods from the shipper.

OVERHEAD BRIDGE CRANE – A shop floor unit consisting of a trolley or crab equipped with a hoisting mechanism, a bridge travelling on wheels fitted to end carriages and supported by track rails, which in their turn are fixed to rail supports provided either at the top of the shop walls or columns.

OVER LOAD PROTECTION DEVICE – Units provided on the hoisting motion that automatically cut out the hoisting motor on jib and tower cranes when the excess of load lifted is not less than 10 per cent of the rated capacity.

PALLET – A standard platform on which material is placed for storage and movement. The platform has an upper and lower flat surface with space between for the forks of an industrial lift truck.

PIGGYBACK (TOFC) – Transportation of highway trailers or demountable trailer bodies on specially equipped cars.

PIN WHEEL PLATFORM – A pallet pattern used for the arrangement of items of unequal length or width.

PIPE LINES – Specialized carriers that transport petroleum or natural gas.

PLATE CLAMP – Handling attachment which is an eccentric clamp used for handling steel sheets and plates in the vertical position.

PNEUMATIC CONVEYING – Method of transporting bulk materials in the form of powder, short fibre and granules over a pipe line as a mixture with air or due to the pressure of air, *e.g.*, suction or vacuum type systems, pressure type systems, combination systems.

PNEUMATIC DRIVE – System in which compressed air is admitted into direct acting cylinders under pressure, the piston rods are linked with the business end.

PORTABLE PLATE – A loading ramp that can be moved to any loading position on the deck.

POSITIONER – A device that orients and positions the part for the next operation.

POWER DRIVE – Drive provided in the form of electric motor, steam engine, internal combustion engine, hydraulic or air motor. Includes combinations such as diesel electric, electro hydraulic, electro pneumatic drives.

POWER SHOVEL – An earth digging machine with an operating member (bucket) which digs soil, carries it over a small distance and dumps it in a pile or onto transporting facilities.

PREFORMED WIRE ROPES – The ropes in which the wires and strands, before being laid, are preformed to fit the shape they carry in the rope.

RACK – A storage place for individual items or palletized loads.

RAIL ROAD CRANE – Crane mounted on a truck meeting railroad requirements and supported by axles of varying number used for load handling and wreckage clearing purposes, carry a power unit which is a combination of a diesel, a generator and a number of electric motors.

REACH OF CRANE – Distance of the load from the axis of the rotary part of the crane, decided by crane stability, decreases with an increase in the load to be handled.

REVERSIBLE PALLET – A pallet with an identical top and bottom deck.

REVOLVING CRANE – A crane which can lift up the load and transfer it to any point of a circle with a radius equal to the reach, being thus suitable for loading the rail road flat cars and lorries, for construction site works etc. *e.g.*, travelling revolving crane, fixed revolving crane.

RIGHT HAND LAY ROPES – Ropes in which the strands bend round to the right.

ROW PATTERN – A pallet pattern used for the arrangement of items of unequal length or width.

SCRAPER – Machine that removes earth slice by slice, transports and places it in an earth structure, or pushes it to a dump and then levels the same.

SCREW CONVEYORS – Installations serving to move materials over a trough by a rotating screw. Also called SPIRAL CONVEYORS.

SEAL – A small metal strip that is placed on a railroad car door when loading is completed.

SELF CLOSING GRAB – A crane grab which can pick up the load automatically but must be released with manual assistance.

SELF LEVELING – Bringing the lift to a stop so that its platform is level with the landing.

SEMI GANTRY CRANES – Hoisting installations in which one pair of legs travels over a track placed on a low foundation while the other pair receives support from an elevated track laid on a trestle or track girder attached to the building.

SHAKING CONVEYOR – Unit which consists of a deck in the form of a trough or pan suspended from, or supported by a stationary frame and the deck is induced to oscillate, causing the material to move along at a pace and in a mode governed by the character of the oscillations, *e.g.*, oscillating conveyors and vibrating conveyors.

SINGLE FACED CORRUGATED FIBRE BOARD – Fibre board with a flat facing of corrugated material on one side.

SINGLE FACED PALLET – A pallet consisting of one deck.

SINGLE WALL CORRUGATED FIBRE BOARD – Fibre board with a flat facing of corrugated material on both sides.

SKID – Similar to pallets except it has no bottom surface.

SKIP HOIST – Conveyor which consists of a bucket moving between guides, with the power of a cable reeved round sheaves and drums which turn through electric drives. Used for vertical movement of coal.

SLIP SHEET – A single sheet of heavy, strong, corrugated fibre board or plastic. Unit load is pulled on it and the forks of the lift truck slide underneath.

SLOTTED STYLE BOX – A box with slots, made from a single piece of fibreboard.

SOFC – A container stacked on a flat car.

SOLID FIBRE BOARD – Two or more layers of fibre board glued together to make a single sheet.

SPAN – Horizontal distance between the axes of crane track rails.

SPEED GUARD – Device which prevents the hook loads from acceleration beyond a set limit. They are usually centrifugal brakes (*e.g.*, centrifugal disc brake, centrifugal internal block brake), hydraulic units and eddy current brakes.

SPIRAL or SCREW CONVEYORS – Unit which employs an endless hellicoid screw rotating in a trough. The movement of the screw gradually moves the material to a higher location.

STABILITY OF A CRANE – The ratio of the righting and tipping moments taken about a tipping axis.

STABILITY COEFFICIENT OF CRANE PROPER – The ratio of the moment produced by the dead weight of all crane components, with the allowance for a track gradient effect facilitating the chance of overturning, taken about a tipping axis to the moment set up by a wind pressure exceeding the safe maximum operating value taken with respect to the same tipping axis.

STABILITY COEFFICIENT OF A CRANE UNDER A LOAD – Ratio of the moment set up by the dead weight of crane components with the allowance of additional forces (wind effect, inertia at starting and stopping the hoisting, slewing and travelling motions, effect of track ground gradient) taken about a tipping axis to the moment due to the working load about the same axis.

STOP OFF CAR – A car that stops at two or more destinations to finish loading or to partly unload.

STOWAGE – Arrangement of freight in the hold of the ship.

STRINGER – Timber that runs the length of a pallet.

TANK TRUCK – A truck in which liquids are transported.

TRACTOR – A crawler or wheel mounted, self propelled vehicle designed to perform agricultural, earthmoving, road making, load carrying and other work.

TRACTOR TRAILER TRAIN – The combination of a puller unit with a train of wheeled dollies.

TRAVELLING CANTILEVER WALL CRANE – Crane arranged to travel over tracks secured to a side wall of the building.

TRAVELLING GANTRY CRANE – A crane which has a rigid four legged gantry travelling over a track laid along the work area, topping the gantry is a rotary part with a jib, a movable counter weight, hoisting and luffing units.

TRUCK – A self powered wheel vehicle used for carrying loads and people.

TRUCK CRANES – Cranes mounted on chassis of standard design or on special purpose chassis with tyred wheels.

TYPE I HAND CRANK – Unit directly linked to the load brake and requiring a constant effort in order to lower the load which descends at a rate corresponding to the rate of crank rotation.

TYPE II HAND CRANK – An arrangement to effect lowering of load at a speed exceeding that of hoisting and finds application where this mode of operation is called for by process requirements.

TWO WAY PALLET – A pallet design that permits forks or hand pallet trucks to enter from two sides only and in opposite directions.

UNITIZED LOAD – A grouping of smaller shipments that are bound together to travel as a single larger unit, also called UNIT LOAD.

UPENDERS AND DUMPERS – Specialized machines used to unload materials handling carriers at a desired location.

VACUUM LIFTER – End of line attachment, a convenient means of handling sheet stock of various kind (steel, non-ferrous metals, glass etc.) along with cartons, boxes etc. To lift a load, the vacuum cup is applied to its surface, and the vacuum generator is set into operation, creating a vacuum in the cup.

VERTICAL LIFT – A ferris-wheel type conveyor that moves goods from the receiving area to the checking area.

WEIGHT OF A BODY – Force which the body exerts on the support due to the force of gravity.

WINCH – Equipment used for hauling, pulling, or hoisting *e.g.*, rope and chain winches, stationary winches, mobile winches.

CHAPTER 14

FRICTION, LUBRICATION AND BEARINGS

ABSOLUTE VISCOSITY – Force per unit area required to move a surface at unit velocity, when it is separated by a fluid of unit thickness from a stationary surface.

ADDITIVES – Chemical compounds used to alter the characteristics of lubricating oils.

ANGLE OF REPOSE – The minimum inclination which a plane can have consistent with the body on it sliding down the plane by the force of gravity. It is the same as the friction angle.

ANTI CORROSION ADDITIVES – Chemical compounds added to lubricating oil to reduce or prevent the chemical reaction of acids (formed by the oxidation of oil) which destroy some bearing materials, such as lead in lead copper bearings, used for crankshafts and connecting rods of engines.

ANTI OXIDATION ADDITIVES – Chemical compounds added to decrease oxidation of the oil. These have a greater affinity for oxygen than does the oil.

API HEAVY DUTY TYPE OIL – Motor oil having oxidation stability, bearing corrosion preventive properties, and detergent-dispersant characteristics necessary to make it generally suitable for use in both high speed diesel and gasoline engines under heavy duty service conditions.

API PREMIUM TYPE OIL – Motor oil having the oxidation stability and bearing corrosion preventive properties necessary to make it generally suitable for use in internal combustion engines where operating conditions are more severe than regular duty.

API REGULAR TYPE OIL – Motor oil generally suitable for use in internal combustion engines under moderate operating conditions.

BALL BEARING – An antifriction bearing where the rolling elements are spherically shaped. Bearing consists of an inner and outer hardened steel races separated by a series of hardened steel balls.

BATH LUBRICATION – Lubrication system in which the bearing contains a space filled with oil, which is in contact with a portion of the journal.

BEARING – The part which transmits the load to the support and in so doing, takes the friction caused by the moving parts in contact. Area of the unit in which the contacting surface of a revolving part rests.

BEARING CAPS – On an engine, caps held in place by bolts or nuts which, in turn, hold bearing halves in place.

BEARING CRUSH – The additional height over a full half which is purposely manufactured in each bearing half. This ensures complete contact of the bearing back with the housing bore when the unit is assembled.

BEARING FAILURE – Failure of a bearing due to continued flexing of the bearing surface from the applied load.

BEARING OIL CLEARANCE – The space purposely provided between the revolving shaft and the bearing in which it rotates. Through this space lubricating oil can flow.

BEARING PRELOAD – Amount of static pressure exerted on a bearing or a set of bearings. Preload is usually adjusted by a threaded collar or shims.

BEARING PRELUBRICATOR – A special tank attached to an airline which supplies oil at a predetermined and maintained pressure to the engine lubricating system when the engine is not operating.

BEARING SPACER – A piece of tubing used between the two wheel bearing inner races to prevent unwanted bearing preload as the axle is tightened.

BEARING SPIN – A type of bearing failure caused by lack of lubrication which overheats the bearing while the crankshaft is still in place.

BEARING SPREAD – A purposely manufactured small extra distance across the parting faces of the bearing half in excess of the actual diameter of the housing bore.

BYPASS FILTER – An oil filter that constantly only filters a portion of the oil flowing through the engine or machine.

BOUNDARY FRICTION – The resistance to relative motion when one solid surface is caused to move tangentially over another, the surfaces being covered only by an adsorbed contamination film.

BOUNDARY LUBRICATION – Type of lubrication in which the two surfaces have between them a more or less complete layer of oil which is only, at the most, a few molecules thick.

BUSHING – A member that takes up space and usually allows movement at the attachment point. A one piece replaceable sleeve placed in a bore to serve as a bearing surface. Bearing for shaft, spring shackle, piston pin etc. A metallic or synthetic lining for a hole.

CARBON RESIDUE – Determined (canradson carbon test) by evaporating under specified test conditions, a known weight of oil and weighing the residue.

CENTIPOISE – A unit of viscosity of a fluid used in figuring pressure drop etc.

CENTRIFUGAL OIL SLINGER – Cup shaped centrifugal oil filter mounted to the end of the crankshaft. As the oil passes through the slinger, centrifugal force removes impurities that are heavier than oil.

COEFFICIENT OF FRICTION – Ratio between the resistance due to friction in the direction of motion and the load carried normal to the plane of motion.

COLLAR BEARING – The collar which may be at any part of the length of the shaft, takes up the thrust produced along the axis of the shaft.

CRANKCASE DILUTION – Dilution of lubricating oil in the oil pan by liquid gasoline seeping down the cylinder walls. Accumulation of unburned gasoline in the crankcase.

CRITICAL SPEED – The limiting or critical speed corresponding to a given pressure is that speed at which surface irregularities may intervene and so lead to seizure.

DETERGENTS – Added to lubricating oils to improve the tendency of the oil to wash or cleanse the surface where oxidation products form, thereby resulting in a marked reduction in ring sticking, particularly in heavy duty service, and a marked improvement in cleanliness of pistons.

DETERGENT OIL – An oil which keeps particles and contaminants in suspension and has ability to neutralize acids resulting from combustion process.

DIRECT BEARING LUBRICATION – An oil injection system which feeds undiluted oil to two stroke cycle engine main bearings and rod big end bearings.

DRY SUMP-LUBRICATION – In this system, oil is gravity fed to supply side of oil pump from the remote oil tank. After the oil has been pumped through four stroke cycle engine, it is returned to the oil tank by return side of oil pump.

FILM LUBRICATION – Type of lubrication in which the two metallic surfaces are separated by a continuous film of liquid oil of measurable thickness which forces itself between them.

FIRE POINT OF OIL – The temperature at which the oil vapour on the oil surface ignites and burns for at least five seconds, when heated gradually, under specified test conditions.

FLASH POINT OF OIL – The temperature at which a momentary flash appears on the heated oil surface when a test flame is applied, under specified test conditions. It is a rough indication of the tendency of the oil to vapourize.

FOOT STEP BEARING – The bearing at the foot of a vertical shaft.

FORCE FEED LUBRICATION – A gear pump takes oil from the sump and delivers it to the distributor ducts which connect with all the main bearings and camshaft bearings, from where the oil reaches the various parts that need lubrication.

FORCE OF FRICTION – The least force acting parallel to the sliding surfaces of the bodies in contact, which will cause one body to slide over the other.

FRICTION – The force which acts between two bodies at their surface of contact so as to resist their sliding on each other, due to roughness of the surfaces. Also called FRICTIONAL FORCES.

FRICTION ANGLE – Angle which the resultant force makes with the normal to the plane when sliding begins. Also called LIMITING ANGLE OF RESISTANCE or LIMITING ANGLE OF REACTION.

FRICTION BEARINGS – Bearings having sliding contact between the moving surfaces. Sleeve bearings, such as those used in connecting rods, are friction bearings.

GEAR OIL PUMP – An oil pump with gear type teeth to move oil.

GREASE – The term is applied to a mixture of mineral oil with fats that have been saponified with an alkali. To this mixture fillers may or may not be added.

HYDRODYNAMIC LUBRICATION – A condition that occurs when a film of oil is constantly maintained between moving parts.

INNER RACE – In a bearing assembly, the part that connects a rotating shaft to the bearing.

JOURNAL – In a shaft, the hardened polished area that is used with a bearing.

LAWS OF FRICTION – The force of friction is (1) directly proportional to the pressure between the surfaces in contact, (2) independent of the extent of the surfaces in contact and (3) independent of the velocity of sliding.

LOAD MOTION – Rolling or sliding motion found at the bearing surfaces.

LUBRICANT – A fluid having two essential properties namely viscosity and oiliness, and used to reduce friction.

LUBRICATING FILM – A thin coating of lubricant (oil) which prevents contact between moving parts.

LUBRICATING SYSTEM – The system in the engine that supplies lubricating oil to the moving engine parts, to prevent actual contact between the moving surfaces.

LUBRICATION – Use of lubricant (oil or grease) to reduce friction.

MAIN BEARINGS – In the engine block, the bearings that support the crankshaft.

MICHELL THRUST BEARING – Type of tilting pad thrust bearing, the thrust supporting surface is divided into a number of segmental pads, which pivot on a radial edge at the back, and so automatically adjust themselves to the required slope.

MULTIGRADE OIL – An oil that retains its viscosity under varying temperature conditions better than a single weight oil.

ML OIL – Oil for light automotive service.

MM OIL – Oil for medium or average automotive service.

MS OIL – Oil for severe automotive service.

MULTI WEIGHT OIL – Type of oil that provides adequate lubrication at both high and low temperatures.

MULTIPLE VISCOSITY OIL – An engine oil which has a low viscosity when cold (for easier cranking and a higher) viscosity when hot (to provide adequate lubrication).

NEEDLE BEARING – Antifriction bearing of the roller type, the rollers are very small in diameter (needle sized). Rollers have a length over four times greater than their diameter.

NEEDLE LUBRICATOR – A needle which rests on a journal and exposed at top to the oil in an inverted glass bottle or reservoir, causes oil to flow slowly onto the journal, due to the vibration set up during shaft rotation.

OIL – A liquid lubricant derived from crude oil used to provide lubrication between moving parts.

OIL CLEANER – The filtering device through which oil passes, which filters dirt and dust from the oil.

OIL CONSUMPTION – Oil pumped into the combustion chamber of an engine by the piston and rings and burnt there in a given time.

OIL CONTROL RINGS – The lower ring or rings on the piston of an engine, designed to prevent excessive amounts of oil from working up into the combustion chamber.

OIL COOLER – A small radiator through which the oil flows to lower its temperature.

OIL DILUTION – Dilution of oil in the crankcase of a SI engine, by leakage of liquid gasoline from the combustion chamber past the piston rings.

OIL FILTER – Unit in the lubrication system through which oil passes to remove any impurities from oil. It may be paper, wire screen or rotor designed to keep oil clean.

OIL LEVEL INDICATOR – The indicator, usually called the dipstick, that can be removed to determine the level of oil in the crankcase of an engine or machine.

OIL PAN – The detachable lower part of the engine made of sheet metal, which encloses the crankcase and acts as an oil reservoir.

OIL PASSAGES – Holes and passages drilled in the block and in the engine parts through which oil flows.

OIL PRESSURE GAUGE – An instrument used to measure and indicate the amount of oil pressure in the lubrication system.

OIL PUMP – In the lubricating system, the device that delivers oil from the oil pan to the various moving engine/machine parts.

OIL PUMP BLEEDING – Removal of air from the supply line and the pump in a two stroke cycle oil injection system.

OIL PUMPING – Passing of oil past the piston rings into the combustion chamber because of defective rings, worn cylinder walls etc.

OIL RESERVOIR – That space in the base of the oil separator where oil is accumulated prior to its return to the pump.

OIL SEAL – A seal placed around a rotating shaft or other moving part, to prevent passage or escape of oil.

OIL SEAL AND SHIELD – Two devices used to control oil leakage past the valve stem and guide into the ports or the combustion chamber.

OIL SPILL RING – Collars or oil throw rings put on the rotating shaft, prevent oil leakage along the shaft, utilizing the increase of centrifugal force with the increase of diameter.

OIL STRAINER – A wire mesh screen placed at the inlet end of the oil pump pickup tube to prevent dirt and other large particle from entering the oil pump.

OUTER RACE – In a roller/ball bearing assembly, the stationary part that supports the rotating load on the bearing.

OXIDIZED OIL – Oil that has been chemically combined with oxygen as a result of excessive heat, oil agitation, and exposure to combustion contaminants.

PAD LUBRICATION – A part of the bearing surface upon which there is no pressure is dispensed with, and its place is taken by a soft pad, which is kept saturated with lubricant.

PICKUP SCREEN – A screen in the oil pan that prevents any large particles from entering the oil pump.

PIVOT BEARING – Pivot portion (flat or conical) which is on the end of a shaft, that takes up the thrust produced along the axis of the shaft.

PLAIN BEARING – A one or two piece assembly consisting of a hard metal backing piece with a softer metal mated to it.

PLUNGER OIL PUMP – An oil pump that uses a piston or plunger and check valves to move oil.

POUR POINT OF OIL – The temperature at which the oil will not flow when cooled under specified conditions, due to formation of wax.

PREMIX – Two stroke cycle engine lubrication is provided by oil, mixed with fuel in the fuel tank.

PRESSURE RELIEF VALVE – A spring loaded valve used to regulate the pressure in the lubrication system.

PRESSURIZED FEED OIL SYSTEM – A type of engine lubricating system that makes use of an oil pump to force oil through tubes and passages to the various engine parts requiring lubrication.

PROFILOMETER – Instrument used to measure the surface irregularities.

RING LUBRICATION – Two loose rings carried by the journal rotation, being driven by frictional contact with the journal, dip into an oil bath and carry oil to the top of the journal so as to flow over the surface of the journal through oil grooves in the bearing, and finally return to the bath below.

ROLLER BEARING – An antifriction bearing consisting of hardened steel inner and outer races, separated by hardened steel rollers, which are two or three times as wide as their diameter.

ROLLING FRICTION – The force resisting the motion when a body rolls on a surface.

ROTARY OIL PUMP – An oil pump which uses an inner and outer rotor to move oil.

SCHIELE'S PIVOT – A pivot bearing on a shaft, designed to give uniform wear in the direction of the axis with uniform pressure, the coefficient of friction being assumed to be constant.

SCRAPER RING – A piston ring that assists in compression and wipes off the excess oil from the cylinder walls.

SEAL – A material, shaped around a shaft, used to close off the operating compartment of the shaft, preventing oil leakage.

SEIZING – The phenomenon in which the metal surfaces adhere and then tear owing to relative motion, caused by the attraction between molecules on opposite sides of the common surface between two elements of a bearing.

SERVICE RATING – For lubricating oils used in engines, a designation that indicates the type of service for which the oil is best suited.

SHIM (bearing) – A strip of copper or similar material used under a bearing cap, for example, to increase bearing clearance.

SKIMMING – Using a machine to remove oil or scum from the surface of the water.

SLUDGE – Accumulation in oil pan, containing water, dirt and oil, sludge is very viscous and tends to prevent lubrication.

SPLASH FEED OIL SYSTEM – A type of engine lubricating system that depends on the splashing of oil for lubrication of moving parts. The engine is enclosed, and the crankcase contains oil into which the cranks dip and splash as they rotate, throwing the oil over the various bearings.

STATICAL FRICTION – Friction at starting from rest, and this is greater than the friction of motion and depends on the hardness of the bodies and the length of time during which they have been in contact.

SUMP – A system for storing oil, either in the crankcase (wet sump) or in a separate tank (dry sump).

SYNTHETIC OIL – Oil made from material other than petroleum.

SYPHON LUBRICATOR – The unit in which the oil stored in a cup above a journal, is delivered slowly to the bearing through a wick which acts as a syphon.

TAPERED ROLLER BEARING – An antifriction bearing where the rolling element is similar to a roller bearing, however it has a taper along its axis, allowing it to accept axial loads in one direction.

THICK FILM LUBRICATION – Lubrication arrangement wherein there is no metal to metal contact. Also called PERFECT LUBRICATION.

THIN FILM LUBRICATION – Type of lubrication that occurs when the condition of bearing operation reduce the thickness of the oil film sufficiently to make the rubbing of the bearing surfaces a near possibility. Oiliness of the lubricant is important in this type of lubrication.

THRUST BEARING – In the engine, the main bearing has thrust faces which prevent excessive endwise movement of the crankshaft.

VALVE STEM SEAL or SHIELD – A device placed on or surrounding the valve stem to reduce the amount of oil which can get on the stem and thereby work its way down into the combustion chamber.

VARNISH – A build up of oxidized oil or fuel.

VISCOSITY – Resistance of oil to flow. The thickness of an oil is determined by its rate of flow, a thick oil has a greater viscosity than a thin oil. Unit of viscosity is centipoise.

VISCOSITY INDEX – A measure of the change of viscosity with temperature of an oil compared with two reference oils having the same viscosity at 210°F one of naphthenic base and the other of paraffinic base.

VISCOSITY INDEX IMPROVERS – Additives added to an oil, to reduce the change of viscosity with increase in temperature.

VISCOSITY RATING – An oil classification based on the time it takes for an amount of oil to flow through a capillary tube, under specified conditions, the higher the rating, the thicker the oil.

WARNING LIGHT – A light on the instrument or control panel that turns on to warn of low oil pressure.

WORK OF FRICTION – Product of the frictional force and the distance through which it moves.

CHAPTER 15

THERMODYNAMICS AND HEAT TRANSFER

ABSOLUTE TEMPERATURE – The temperature of a substance measured above absolute zero.

ABSOLUTE ZERO TEMPERATURE – Temperature at which all molecular motion ceases, according to the kinetic theory of gases. A point which has been determined on the thermodynamic scale (by theoretical considerations) beyond which a further decrease in temperature is inconceivable. This is equal to – 459.6° on the fahrenheit scale and – 273.1° on the centigrade scale.

ABSOLUTE TEMPERATURE SCALES – Used for calculating changes in refrigerant vapour pressures.

ADIABATIC EXPANSION or COMPRESSION – Expansion or compression where the temperature rises during compression and falls during expansion without any loss of heat to the cylinder walls or absorption of heat from the walls.

ADIABATIC HEAT DROP – The heat energy released and theoretically capable of transformation into mechanical work during the adiabatic expansion of unit weight of steam or other vapour or gas.

ADIABATIC SYSTEM – is the system which is insulated from its surroundings. In this system no heat transfer takes place *i.e.*, either into the system or out of the system. It can however exchange work with the surroundings.

AIR – A gas consisting principally of a mechanical mixture of 23.2 per cent (by weight) of oxygen, 65.5 per cent of nitrogen and 1.3 per cent of argon.

AIR STANDARD EFFICIENCY – The thermal efficiency of an internal combustion engine working on the appropriate air standard cycle.

AMBIENT TEMPERATURE – Temperature (usually of the air) surrounding the operating equipment.

APPLIED THERMODYNAMICS – Also called engineering thermodynamics deals with special applications such as energy transfer as power generation, refrigeration and, compression and expansion of gases and fluids.

ATKINSON CYCLE – A working cycle for internal combustion engines, in which the expansion ratio exceeds the compression ratio, more efficient than the Otto Cycle, but mechanically impracticable.

ATMOSPHERIC LINE – A datum line drawn on an indicator diagram by allowing atmospheric pressure to act on the indicator piston or diaphragm.

ATMOSPHERIC PRESSURE – Force exerted by the weight of the atmosphere on every point with which it is in contact. It is generally taken as 1.03 kscm at sea level.

AVAGADRO'S LAW – Equal volumes of different gases at the same temperature and pressure contain the same number of molecules.

AVAILABLE ENERGY – That portion of a quantity of heat which could be transformed into work by means of a reversible engine.

BASIC UNITS – are length, mass, time, temperature and angle.

BLACK BODY – A body which absorbs all the radiation falling on it *i.e.*, has a non-reflecting surface. A black body emits the maximum amount of radiation possible at a given temperature, and the amount is proportional to the fourth power of the absolute temperature.

BOILING POINT – The temperature at which a liquid boils for any given surrounding atmospheric pressure. Now the saturation pressure of the vapour equals that of the atmosphere.

BOUNDARY – is a real physical surface or an imaginary surface enclosing some matter. The boundary may be a fixed one or a varying one.

BOYLE'S LAW – The absolute pressure of a gas will vary inversely as the volume, if the temperature remains constant. Or conversely, the volume will vary inversely as the absolute pressure, if the temperature remains constant.

BRAKE HORSE POWER – Useful power available from an engine. Also called SHAFT HORSE POWER.

CALORIE – The heat per unit weight, one gram, required to raise the temperature of water through one degree centigrade.

CELSIUS – The scale of changes of temperature which uses 0 degree as the freezing point and 100 degree as the boiling point for water at standard pressure.

CHARLE'S LAW – At constant pressure, the volume of a gas is proportional to its absolute temperature. At constant volume, the pressure is proportional to its absolute temperature.

CLOSED SYSTEM – The system which will have boundaries across which both heat and work can penetrate, but no mass will be permitted to cross them.

C_p – Specific heat at constant pressure-Heat to be supplied to raise the temperature of 1 kg of gas through 1°C, the pressure being kept constant (in other words external work is done).

C_v – Specific heat at constant volume-Heat to be supplied to raise the temperature of 1 kg of gas through 1°C, the volume being kept constant (in other words no external work is done).

COEFFICIENT OF EXPANSION – The fractional increase in length or volume per degree rise in temperature.

COEFFICIENT OF LINEAR EXPANSION – Amount of expansion per unit length, per degree rise in temperature.

COEFFICIENT OF PERFORMANCE OF A HEAT PUMP – COP_{HP} = Heat added to the hot body/work supplied.

COEFFICIENT OF PERFORMANCE OF A REFRIGERATOR – COP_{Ref} = Heat removed from cold body/work supplied.

COEFFICIENT OF THERMAL CONDUCTIVITY – The quantity of heat that will flow across a unit area in unit time if the temperature gradient across this area is unity.

COLD – A comparative lack of heat, indicating chillness.

COMPRESSED SOLID – is a solid at a temperature below its saturation temperature.

COMPRESSIBILITY FACTOR, Z – is the factor introduced to modify the ideal gas equation Pv = RT, and to describe the behaviour of a real gas. Z = Pv/RT.

CONDENSATE – The liquid formed by the condensation of a gaseous substance.

CONDENSATION – The change of state of a substance from the gaseous to the liquid form.

CONDENSING – The process of giving up latent heat of vapourization in order to liquefy a vapour.

CONDUCTION – Transfer of heat from one part of a material to another or to a material with which it is in contact.

CONDUCTIVITY – The relative value of a material, as compared with a standard, in affording a passage through itself or over its surface for heat.

CONSTANT VOLUME PROCESS – is one wherein a gas is heated (or cooled) in a fixed enclosed space (no change in volume occupied by the gas). There will be no workdone by the gas. The whole heat supplied will be stored in the form of internal energy.

CONSTANT PRESSURE PROCESS – Also called isobaric process. Heat supplied to a system exhibits as the change in enthalpy.

CONSTANT TEMPERATURE PROCESS – Also called isothermal process. There is no change in temperature and hence internal energy and enthalpy remain constant. Heat supplied = work done.

CONVECTION – Passage of heat from one point to another by means of a gravity fluid circulation due to changes in density resulting from picking up and giving up heat. Also transfer of heat to or from a fluid (liquid or gas) flowing over the surface of a body.

COSMIC RADIATION – Radiation of many sorts, but mostly atomic nuclei (protons) with very high energies, originating outside the earth's atmosphere.

COUNTER FLOW HEAT EXCHANGER – A heat exchanger in which the warm substance flows in the opposite direction to the flow of the cool substance.

CRITICAL STATE OF A SUBSTANCE – is that state at which liquid and vapour coexist in equilibrium. At critical state, latent heat of evaporation becomes zero.

CRITICAL PRESSURE – The critical pressure of a vapour is the pressure required to liquefy it at the critical temperature and is the highest pressure on the temperature -pressure graph for saturated vapour.

CRITICAL TEMPERATURE – Temperature of the vapour above which no pressure, however high, will produce liquefaction.

CRITICAL VELOCITY – The velocity above which fluid flow is turbulent.

CYCLIC PROCESS – is a process (or a series of processes) which returns the system to the state it was before the process began.

DALTON'S LAW OF PARTIAL PRESSURES – At a common temperature, a mixture of gases will exert on the sides of the vessel a total pressure equal to the sum of the pressures which each constituent would exert separately if it alone occupied the vessel.

DERIVED UNITS – are those derived from basic units, *e.g.*, force, work, density etc.

DIMENSIONS – are the properties by which the physical quantity of an object may be described.

DRY BULB TEMPERATURE – The actual temperature of air, as opposed to wet bulb temperature.

DRY STEAM – Steam containing no moisture. It may be either saturated or superheated steam.

DRY SATURATED STEAM – Saturated steam, as generated from water, that contains no moisture in suspension.

EFFICIENCY – Ratio of the useful work performed by a prime mover to the energy expended. In other words the output divided by the input.

ENERGY – Stored work, that is the ability to do work.

ENTHALPY – Name given to the total heat in the fluid at any temperature.

ENTROPY TEMPERATURE DIAGRAM – In thermodynamics, the base of a heat diagram, the area of which is heat units and the altitude of which is absolute temperature.

EQUIVALENT EVAPORATION – Amount of water in kg. that would be evaporated from water at 100°C into steam at 100°C and 1.03 kscm, by the heat put into steam actually evaporated in one hour by 1 kg of fuel.

EVAPORATION – Process by which a liquid changes into a vapour as a result of absorbing heat.

EVAPORATIVE CONDENSER – Combines the principles of forced circulation convection currents with the ability of a vapourizing liquid to absorb heat.

EXTERNAL LATENT HEAT – When vapourization takes place, the amount of heat required because of the work in pushing back the atmosphere to make room for the steam.

EXTRINSIC PROPERTIES – Also called extensive properties, are those properties which are dependent on the mass of the system. Examples are volume, weight and total energy.

FACTOR OF EVAPORATION – A quantity which when multiplied by the amount of steam generated at a given pressure from water at a given temperature, gives the equivalent evaporation from and at 100°C.

FLOW WORK – The product PV (pressure and specific volume) represents flow work in a steady flow system.

FORCED CONVECTION – Circulation of the fluid is made positive by some mechanical means such as a pump for water or a fan for hot gases.

FORCED DRAFT COOLING TOWER – Cools water by mechanically forcing air through the water spray in the tower.

FREE EXPANSION – is a process wherein a fluid from a pressure chamber expands into a vacuum chamber through an orifice of large dimensions.

FREEZING POINT – The temperature at which water turns into ice (0°C at sea level).

FROM AND AT 100°C – In boiler operation, it is an evaporation that would be the equivalent of the actual evaporation when the feed water enters the boiler at 100°C and steam is formed at 100°C, at standard atmospheric pressure.

FROSTING EVAPORATORS – Those evaporators which always operate at temperature below 0°C.

FUNDAMENTAL DIMENSIONS – Internationally accepted units are time, length, mass, force and temperature.

GAUGE PRESSURE – Pressure above or below atmospheric pressure.

GAY LUSSAC LAW – The volume of a gas will vary directly as the absolute temperature, if the pressure remains constant.

GASES – are substances in which the molecules move freely and exist in an unbound state. Gases can be compressed easily because of the large empty space available between the molecules.

HEAT – A form of energy and is due to the motion of the molecules of which all substances are composed. Unit of heat is kilocalorie.

HEAT OF COMPRESSION – Increase of temperature brought about by compression of air or air fuel mixture.

HEAT OF CONDENSATION – The heat that is removed per kilogram of vapour to cause it to condense. It has the same numerical value as the heat of vapourization.

HEAT ENGINE – It is an energy transformer. It transforms heat energy into mechanical energy and render the latter available for doing useful work.

HEAT PUMP – is a device which extracts heat from low temperature surroundings and sends it to a high temperature body, while operating in a cycle. Heat pump maintains a body or system at a temperature higher than the temperature of the surroundings, with the work supplied to it.

HEAT SINK – A means for disposing of unwanted heat, usually by using it to increase the temperature of water, which is then run to waste.

HEAT SOURCE – Supplier of heat to the working agent of a heat engine-a fraction of the heat supplied being changed into work.

HEAT TRANSFER – Movement of heat energy from one place to another (warmer to cooler portion).

HETEROGENEOUS SYSTEM – is a system which is made of more than one phase.

HOMOGENEOUS SYSTEM – is a system consisting of a single phase.

HYPERBOLIC PROCESS – is one in which a gas is heated in such a way that at any instant its pressure multiplied by its volume remains constant.

IDEAL GAS – is one which will obey all the gas laws, under all conditions of temperature and pressure. For an ideal gas, the internal energy and enthalpy are functions of temperature alone.

INDICATED HORSE POWER – The actual power developed or used within a cylinder as calculated from the indicator diagram.

INERT GAS – A gas which under normal conditions does not react with or combine with other substances for example, argon, helium, neon, krypton, xenon.

INTERNAL ENERGY – is the energy arising from the motion and from configuration of the internal particles (atoms and molecules). It is the energy stored in the gas and is used for raising its temperature. $v = C_v T.$, $\Delta_u = C_v (T_2 - T_1)$.

INTRINSIC PROPERTIES – also called intensive properties, are those properties which are independent of the mass of the system. Example are pressure, temperature and density.

ISENTROPIC PROCESS – which is called reversible adiabatic process is one in which no heat is supplied and work done is at the expense of internal energy. Law followed is Pv^{γ} = constant.

ISOLATED SYSTEM – is a special case of the closed system which will not be influenced by the surrounding. In this system, no mass transfer, no energy transfer occur. Work and heat cannot cross the boundary of the system.

ISOTHERM – The line drawn through points or areas of similar temperature.

ISOTHERMAL EXPANSION or COMPRESSION – The expansion or compression of a gas at constant temperature, that is with the gas temperature remaining the same during the process.

JOULE'S LAW OF INTERNAL ENERGY – The internal energy of a given quantity of a gas depends only on the temperature of the gas. As temperature changes, the internal energy also changes.

KELVIN PLANK STATEMENT – It is impossible to construct an engine undergoing a cyclic process, which will convert all the heat supplied to it into an equivalent amount of work.

KILO PASCALS (kPa) – Measurement of pressure in the metric system 1 kilo pascal is approximately equal to 6.895 pounds per square inch.

KINETIC ENERGY – Energy due to momentum, that is the energy of a moving body, which is equivalent to saying, dynamic inertia. The kinetic energy of a moving body is the work which the body is capable of performing against a resistance before it is brought to rest, that is, it equals the work which has brought it from its state of rest to its actual velocity.

LATENT HEAT – The quantity of heat required to change the state or condition under which a substance exists without changing its temperature.

LATENT HEAT OF FUSION – The amount of heat which must be added to one kilogram of material to change its state from a solid to a liquid or which must be subtracted from one kilogram of a liquid to change it to a solid.

LATENT HEAT (internal) – The amount of heat that the water will absorb (latent heat minus heat equivalent of work of evaporation), at the boiling point without a change in temperature.

LATENT HEAT OF VAPOURIZATION – Amount of heat to be added to (or subtracted from) one kilogram of the refrigerant to cause it to vapourize (or condense). Also, the amount of heat energy in a gas which is in addition to that found in the liquid at the same temperature.

LAW OF CONSERVATION OF ENERGY – States that energy can neither be created nor destroyed. However, it can be stored in various forms and can be transferred from one form to another. In other words, the total energy of a body or substance always remains constant.

LIQUIDS – are substances in which the molecules are loosely bounded. A definite volume is maintained by the bond forces, but can change the shape according to the shape of the container.

MATTER – General name for all material substances, gaseous, liquid or solid forming the earth and its surrounding atmosphere.

MEAN EFFECTIVE PRESSURE – The difference between the mean forward pressure and the mean back pressure acting on a moving piston during a cycle of operation.

MECHANICAL CYCLE – is one wherein during the various processes, the properties of the working fluid change.

MECHANICAL EFFICIENCY – Ratio of mechanical energy available at the brake or flywheel to the mechanical energy put into the moving mechanism at the piston of the engine.

MECHANICAL EQUIVALENT OF HEAT – 1 kcal is equal to 1427 Mtkg of mechanical energy.

MOLE OF GAS – Quantity of gas, the weight of which is equal to the molecular weight in pounds.

MOLAR HEAT – is the quantity of heat required to raise the temperature of one mole of a substance through one degree.

MOLLIER DIAGRAM – A graphical representation of the properties of a fluid in which enthalpy is plotted as abscissa and pressure as ordinate.

NATURAL CONVECTION – Circulation or the transfer of fluid due to a difference in density resulting from temperature changes.

NATURAL DRAFT COOLING TOWER – Cooling tower which cools water by moving air at low velocities through the tower, without the aid of any mechanical means.

NON FLOW PROCESS – It is the one in which there is no mass interaction across the system boundaries during the process. This occurs in a closed system.

NTP – The normal temperature and pressure refer to the conditions of temperature 0°C (273 K) and pressure of 760 mm of mercury.

OPEN SYSTEM – It has no closed boundary, but has one or more openings. Fluid may enter or leave the system, it undergoes thermodynamic process, while moving through the system.

PARALLEL FLOW HEAT EXCHANGER – A heat exchanger in which the warm and cool fluids flow in the same direction but separated from one another.

PERFECT GAS – Gas which behaves in accordance with the gas law PV = MRT, where P-absolute pressure, V-volume, M-mass, T-absolute temperature and R-constant for the particular gas depending on its molecular weight.

PERPETUAL MOTION MACHINE OF FIRST KIND – is a machine which will give continoues work without taking any energy from other system or surrounding.

PERPETUAL MOTION MACHINE OF SECOND KIND – is a machine which abstracts heat continuously from a reservoir and converts it completely into work or 100% efficient engine.

PERPETUAL MOTION MACHINE OF THIRD KIND – is applied to devices that, once set in motion, continue in motion for an indefinitely long time without slowing down.

PHASE OF A SUBSTANCE – It refers to a quantity of matter which is homogeneous or uniform throughout its physical structure and chemical composition.

PHASE DIAGRAM – is a pressure temperature diagram showing more than one of the saturation lines (liquid-vapour, liquid-solid, solid-vapour and other of a pure substance).

PATH FUNCTIONS – depend on which path is followed between the states 1 and 2. Examples are heat, work, internal energy and entropy.

POINT FUNCTIONS – are the properties of a system at a state.

POLYTROPIC PROCESS – follows the law Pv^n = constant. In this process, heat is supplied in such a way there is some work done by the gas (*i.e.*, during expansion) and there is also a change in internal energy.

PROCESS – It is the transformation of a system from one state to another.

PROPERTY OF A FLUID – is its characteristic such as pressure, volume, temperature, internal energy, enthalpy and entropy which are measurable directly or indirectly.

POTENTIAL ENERGY – Energy possessed by a substance by virtue of its position with respect to a datum.

POWER – The rate at which work is done, *i.e.*, 1 HP = 4500 Mtkg/ minute.

PRESSURE – As defined by Rankine-A force of the nature or a thrust, distributed over a surface measured as kg/sqcm.

PURE SUBSTANCE – is any material of single chemical structure or of homogeneous and invariant chemical structure.

QUALITY OF STEAM – Refers indirectly the amount of water or unevaporated moisture in steam.

QUASI-STATIC PROCESS – is a process which is carried out in such a manner that at every instant the system departs only infinitesimally from the thermodynamic equilibrium state.

R – An experimentally determined constant which is equal to the mechanical work done by the expansion of unit weight of a perfect gas at a constant pressure while heat is added to increase its temperature through one degree.

RADIATION – Continuous form of interchange of heat energy from one body to another by means of electromagnetic waves without causing a change in the temperature of the medium between the two bodies involved. These energy waves may be reflected, penetrate the material or be absorbed.

REAL GAGES – are those that deviate from the characteristic gas relation, $Pv = RT$.

REFLECTIVITY – Ability of a material to reflect radiant heat.

REFRIGERATOR – Maintains a body at a temperature lower than that of the surroundings while operating in a cycle. External work is supplied for realizing it.

RENAULT'S LAW – The two specific heats of a gas C_v and C_p do not change with the temperature and pressure.

REVERSIBLE PROCESS – Any process performed so that the system and all its surroundings can be restored to their initial states by performing the process in reverse.

RUDOLF CLAUSIUS STATEMENT – Heat can flow from a hot body to a cold body unaided, but it cannot flow from a cold body to a hotter body without the expenditure or supply of mechanical work.

SATURATION TEMPERATURE – Temperature of the liquid surface corresponding to the pressure of the vapour in contact with it.

SATURATED PHASE – is any phase of a substance existing under saturated conditions, wherein two or more phases of a pure substance can exist together in equilibrium.

SATURATED STEAM – Steam, as it is generated from water, just barely on the steam side of the fence.

SATURATED VAPOUR – Vapour whose temperature and pressure are in accordance with the temperature vapour pressure relationship for the particular substance. Vapour in contact with its liquid is saturated.

SECONDARY DIMENSIONS – are those quantities which are described using primary dimensions.

SEMI-PERFECT GAS – is one which follows the ideal gas relations with specific heats being functions of temperature.

SENSIBLE HEAT – Heat energy that causes a change of temperature of a solid, liquid or gas, changes the speed with which molecules move. When unit weight of a substance is heated by one degree to gain, the sensible heat is equal to the specific heat.

SENSIBLE HEAT OF WATER – Heat added to one kg of water from 0°C to the boiling point at the given pressure. Also called ENTHALPY OF WATER.

SOLIDS – are substances in which the molecules are closely packed and may not move freely. Molecules can only vibrate about a mean position. Very large force is required to change the shape of a solid substance.

SPECIFIC HEAT – The ratio of the amount of heat which will raise the temperature of a given weight of a substance by one degree to the amount of heat which will produce the same raise in temperature in the same weight of water.

SPECIFIC HEAT OF SUPERHEATED STEAM – Amount of heat required to raise the temperature of superheated steam, at constant pressure by 1°C.

SPECIFIC PROPERTY – is the value of any extensive property divided by the mass of the system. This is an intensive property, *e.g.*, specific volume, specific weight.

SPHEROIDAL STATE – The condition of a liquid, as water, when being thrown on a highly heated metal surface, it rolls about in spheroidal drops or masses, at a temperature several degrees below ebullition and without actual contact with the heated surface. This phenomenon is due to the repulsive force of heat and the intervention of a cushion of vapour.

STATE OF A SYSTEM – refers to the unique condition of the system at any particular moment. The state is indicated by the properties such as temperature, volume, pressure, internal energy, enthalpy and entropy. These are point functions.

STEADY FLOW SYSTEM – is an open system in which the rate of mass transfer and energy transfer remain constant with respect to time.

STEAM – The vapour of water. It is a colourless expansive invisible gas.

STEAM TABLES – Tables containing values of various properties of saturated steam such as boiling point, specific volume, sensible heat, latent heat and total heat calculated for a wide range of pressures.

STEADY FLOW PROCESS – is a flow process wherein the conditions within the control volume do not vary with time.

STORED ENERGY – is the energy which is contained by the medium within the system boundaries. Examples are potential energy and internal energy.

STP – The standard temperature and pressure refer to the conditions of temperature 15°C (288 K) and pressure 760 mm of mercury.

SUBCOOLED LIQUID – is a liquid existing at a temperature lower than its saturation temperature (or, in other words, at a pressure higher than its saturation pressure).

SUPERHEAT – Addition of heat to a fluid after it has completely vapourized. In this situation, temperature increases but pressure does not.

SUPERHEATED VAPOUR – Vapour removed from contact with its liquid and at a temperature higher than that which corresponds to its pressure as indicated by the temperature – pressure vapour relationship for that particular substance.

SUPERHEATED STEAM – Steam at a temperature higher than the saturation temperature for the given pressure.

SURROUNDING – is everything outside the boundary which may influence the behaviour of the system. Transfer of mass and energy may take place between the system and boundary.

SYSTEM IN EQUILIBRIUM – means it does not undergo any change on its own accord.

SYSTEM IN CHEMICAL EQUILIBRIUM – is an isolated system which may contain substances that can react with one another chemically, but all these chemical reactions cease after a certain time.

SYSTEM IN MECHANICAL EQUILIBRIUM – is an isolated system, when left to itself, may change with time all its properties such as pressure, elastic stress etc., but these changes may cease after a certain time.

SYSTEM IN THERMAL EQUILIBRIUM – is an isolated system, wherein there may be variations of temperature from point to point, but these variations vanish after a certain time.

SYSTEM IN THERMODYNAMIC EQUILIBRIUM – is one which is in mechanical equilibrium, thermal equilibrium and chemical equilibrium.

TEMPERATURE – A measure of the intensity of heat in a substance with reference to a fixed reference point and of its ability to pass its heat into anything at a lower temperature than itself.

THERMAL CONDUCTIVITY – Ability to a material to allow the passage of heat. Stated as the number of kilocalories that can flow in one hour through a block of material, one square meter in area and one meter thick with one degree centigrade difference in temperature between the opposite surfaces.

THERMAL GRADIENT – The slope of the temperature curve produced by measuring the temperature at various points across a body or heated section.

THERMODYNAMICS – Study of energy, heat and work, properties of media employed and the processes involved.

THERMODYNAMIC SYSTEM – is any space or any matter or group of matter within a prescribed boundary, to which attention is directed for the purpose of analysis.

THERMODYNAMIC CYCLE – is said to occur when a working fluid of a system undergoes a number of operations and processes which take place in a certain order and finally return the fluid to the initial state itself.

THERMODYNAMICS – ZEROTH LAW – That two systems having equal temperatures with a third system also have equal temperatures with each other.

THERMODYNAMICS – FIRST LAW – In ordinary processes, matter can neither be created nor destroyed. Energy can take different forms but cannot be created or destroyed.

THERMODYNAMICS – SECOND LAW – There is a definite limit to the amount of mechanical energy that can be obtained from heat energy.

THERMOELECTRIC INDICATORS – Temperature measuring instruments which operate on the principle that minute quantities of electric current may be produced by heating two dissimilar metals which are joined at one end.

THERMOSTAT – Operating control which reacts to temperature.

THROTTLING PROCESS – is one that occurs when a gas or vapour is expanded through an aperture of minute dimensions, such as a slightly opened valve or a narrow throat. In this process, no heat transfer takes place, no work is done.

TOTAL HEAT – Sum of the sensible heat and latent heat. Also called enthaply of steam or vapour.

TRANSIT ENERGY – is that energy which crosses the system boundaries. Examples are heat, work, velocity energy and electrical energy.

TRIPPLE POINT – The temperature and pressure at which a substance can exist simultaneously in solid, liquid and vapour forms.

TUBE AND FIN RADIATOR CORE – One type of radiator core, consisting of tubes to which cooling fins are attached, water flows through the tubes.

ULTRAVIOLET RADIATION – Radiation of short wave length or high frequency, just below the X-ray range, which come from the SUN and the electric arc. These may cause eye damage and skin burns unless the correct precautions are followed.

UNIVERSE – is the system and the surrounding put together.

UNSTEADY FLOW SYSTEM – is an open system in which the rate of mass transfer and energy transfer vary with respect to time.

VACUUM – A space devoid of matter, that is, a space in which the pressure is zero absolute.

VAPOUR – A gaseous condition of a substance near its point of liquefaction.

VAPOURIZATION – Change of state from liquid to vapour or gas.

VAPOUR PRESSURE – Pressure exerted by the vapour of any particular liquid on the containing vessel, as a result of the molecules near its surface freeing themselves from the attraction of their neighbours and flying off into space. Its magnitude depends solely on the temperature of the liquid surface.

VOLUMETRIC EFFICIENCY – The ratio of the volume of gas actually pumped by a compressor or pump to the theoretical displacement of the piston of the compressor or pump or reciprocating IC engine.

WEIGHT – It refers to the force exerted by gravity on the given mass. Weight is proportional to mass, but the proportionality factor is different at different locations.

WET STEAM – Steam containing intermingled moisture, mist or spray.

WORK – The overcoming of resistance through a certain distance by the expenditure of energy.

WORKING AGENT – Substance in a heat engine which alternately takes in and rejects heat, expanding and contracting when doing so, and overcoming the resistance opposing these changes.

ZEROTH LAW OF THERMODYNAMICS – States that if two systems are separately in equilibrium with a third system, then the two systems will be in thermal equilibrium with each other.

CHAPTER 16

FUELS AND COMBUSTION

ACCELERATOR – Device for rapid control of the speed of an engine, for quick opening and closing of the throttle which regulates the quantity of air fuel mixture into the engine cylinder.

ACHESON GRAPHITE – That made from coke in an electric furnance.

ACTIVATED CARBON – A highly absorbent form of carbon used to remove odors and toxic substances from gaseous emissions or to remove dissolved organic matter from waste water.

ADDITIVE – A substance added to fuel, or oil or grease which improves the properties of the same.

ADIABATIC FLAME TEMPERATURE – The maximum possible temperature attained by the products of reaction, when the reaction goes to completion and all the heat released is used to heat up the products.

AERATION TEST BURNER – Apparatus by which the combustion characteristics of commercial gases can be correlated and calibrated.

AEROSOL – A particle of solid or liquid matter that can remain suspended in the air because of its small size. Particulates under 1 micron in diameter are called aerosols.

AFTER BOIL – Boiling of the fuel in the carburettor or coolant in the engine immediately after the engine is stopped.

AFTER BURNING – In an internal combustion engine, the persistence of the combustion process beyond the period proper to the working cycle, *i.e.*, into the expansion period.

AFTER BURNER – In an automobile engine, a type of exhaust manifold that burns the hydrocarbon and carbon monoxide remaining in the exhaust gas.

AIR ASPIRATOR SYSTEM – An air injection system using a valve opened and closed by pulses in the exhaust system.

AIR FUEL MIXTURE – Air and fuel travelling to the combustion chamber after being mixed by the carburettor.

AIR FUEL RATIO – Ratio (by weight) between air and fuel that makes up engine fuel mixture.

AIR INJECTION SYSTEM – A system which injects air into the exhaust manifold or thermal reactor so that the combustion of the carbon monoxide and unburned hydrocarbons in the exhaust can be completed.

AIR MASS METERING – In some fuel injection systems, fuel metering is controlled primarily by engine speed and the amount of air actually entering the engine.

AIR POLLUTION – Contamination of earths atmosphere by various natural and man made pollutants such as smoke, gases, dust etc.

ALCOHOL – Volatile liquid fuel consisting wholly or partly, of alcohol, able to withstand high compression ratios without detonation.

ANTHRACITE COALS – Slow burning coals which yield very little ash, moisture and less than 10 per cent volatiles, generally used in closed stoves.

ANILINE POINT – The lowest temperature at which an oil is completely miscible with an equal volume of aniline.

ANTIBACKFIRE VALVE – Valve used in air injection reaction cexhaust emission control system to prevent backfiring during the period immediately following sudden decleration.

ANTIKNOCK – In engine fuels, that property which opposes knocking *i.e.*, autoignition.

ANTIKNOCK COMPOUND – An additive put into gasoline to suppress knocking or detonation *e.g.*, Tetra ethyl lead.

ANTIPERCOLATOR – Device for venting vapours from main discharge tube, or well, of a carburettor.

APIEZON OILS – The residue of almost zero vapour pressure left by vacuum distillation of petroleum products.

API GRAVITY – The American Petroleum Institute (API) has established the formula for calculating the specific gravity of a fuel or oil as Degree API= ((141.5/specific gravity at 60/60 degree F)–131.5). The symbol 60/60 degree F is interpreted as the ratio of the weight of a given volume of oil at 60 degree F to the weight of the same volume of water at 60 degree F.

ASH – An inorganic non-combustible residue obtained by combustion of an oil or fuel in the presence of air.

ASH AND SLAG – Impurities that do not burn and usually troublesome elements in coal fired boilers.

ASH FREE BASIS – When fuels are delivered on an ash free basis, it means that the percentage of the ash has been deducted and the other constituents have their percentages recalculated on 100 per cent total without the ash.

ATMOSPHERIC GAS BURNER SYSTEM – A natural draught burner injector, in which the momentum of a gas stream projected from an orifice into the injector throat inspirates from the atmosphere a part of the air required for combustion.

ATOMIZATION – The spraying of a liquid through a nozzle so that the liquid is broken into a very fine mist.

ATOMIZER – A nozzle through which oil fuel is sprayed into the combustion chamber of an oil engine or boiler furnace. It breaks up the fuel into a fine mist so as to ensure good dispersion and combustion.

AUTOIGNITION – The self-ignition or spontaneous combustion of a fuel when introduced into the heated charge in the cylinder of a compression ignition engine.

AUTOMATIC CHOKE – A carburettor choke device (valve) that automatically positions itself in accordance with the carburettor needs or engine temperature.

BACKFIRE (exhaust system) – Passage of unburned air fuel mixture into the exhaust system where it is ignited by some hot spot and causes a loud explosion.

BALANCED DRAFT – A boiler using both forced draft fan and induced draft fan, can be regulated and balanced in the amount of air and flue gases handled so that the furnace pressure is almost atmospheric.

BAGASSE – A fuel produced as a by product of the abstraction of juice from sugar cane. The dried cane (fibrous residue) is usually fed into a specially designed furnance by means of overfeed stokers.

BENCH – The name applied to a complete plant for the manufacture of coal gas. Also called RETORT BENCH.

BENZOL – Crude benzene, used as a motor spirit, generally mixed with petrol, and valued for its antiknock properties.

BIOGAS – Obtained by fermentation in the sewage disposal system, or by fermentation of cattle waste, farm waste etc.

BIOSPHERE – The portion of earth and its atmosphere that can support life.

BLAST FURNANCE GAS – A gas of low calorific value, a by product of iron smelting due to burning of coke in the furnance with limited air, used for preheating the blast, for steam raising etc. It may contain up to 30% carbon monoxide.

BLAST MAIN – The main blast air pipe supplying air to a furnace.

BLOW BY – Leakage of unburned air fuel mixture and some burned gases past the piston rings into the crankcase during the compression and combustion strokes.

BLOW TORCH EFFECT – In gas or oil burning furnaces, when the flame impinges on any surface, such as a tube or refractory wall, that surface is burned as by a blow torch. This is a combustion condition to be avoided as destructive to the surface.

BLUE WATER GAS – A mixture of approximately equal proportions of carbon monoxide and hydrogen made by passing steam over incandescent coke in special generators.

BOILING POINT – The temperature at which a liquid begins to boil.

BOMB CALORIEMETER – An apparatus used for determining the calorific values of fuels. The bomb consists of a thick walled steel vessel in which a weighed quantity of fuel is ignited in an atmosphere of compressed oxygen. The bomb is immersed in a known volume of water, from the rise of temperature of water the calorific value is calculated.

BOTTLED GAS – LPG (**liquefied petroleum gas**) compressed into strong metal containers. Gas when confined in tank, under pressure, is in the liquid form.

BRIQUETS – Coherent masses of uniform size made by the application of pressure to any powdery material placed in a suitable mould with or without a binder.

BUTANE – A hydrocarbon gas formed synthetically, by the action of zinc or ethyl iodide. Petroleum gas, that is liquid, when under pressure. Often used as engine fuel in trucks.

CALORIE – The amount of heat required to raise one gram of water through 1°C *i.e.*, from 17 to 18°C. Calorie is unit of heat.

CALORIFIC INTENSITY – The maximum flame temperature attained when the fuel is burnt.

CALORIFIC VALVE – The heat value of a fuel, expressed in either BTU per pound or CHU per pound or kilocalories/kg. The amount of heat produced by burning unit weight of fuel.

CALORIEMETER – Measuring instrument used to determine the amount of heat produced when a substance is burned, also friction and chemical change produce heat.

CARBON – One of the non-metallic elements constituting fuel and lubricating oil.

CARBON DEPOSIT – A black, hard or soft deposit left on engine parts by the combustion of fuel. Carbon forms on pistons, rings, plugs, valve heads etc., inhibiting their action.

CARBONDIOXIDE – A colourless, odourless gas which results when hydrocarbon or carbon is burned completely.

CARBONIZE – Building up of carbon on objects such as spark plug, piston head etc., of an engine.

CARBON MONOXIDE – A colourless, odourless, tasteless, poisonous gas found in engine exhaust. Formed when carbon or hydrocarbons are burned incompletely.

CARBURETED WATER GAS – An artificial gas formed by passing steam through a bed of glowing coke and thereafter enriching the gas so formed with petroleum vapour.

CATALYTIC CONVERTER – A muffler like device for use in an exhaust system that converts harmful gases in the exhaust into harmless gases by promoting a chemical reaction between a catalyst and the pollutants.

CETANE NUMBER – Rating of ignition quality or performance characteristic of diesel fuel. A high cetane number fuel ignites more easily at lower temperature than a low cetane number fuel.

CHARCOAL – Product obtained by heating wood out of contact with air.

CHARCOAL CANISTER – A container filled with activated charcoal used to trap gasoline vapour from the fuel tank and carburettor while the engine is off.

CHEMICAL CHANGE – A change which alters the composition of the molecules of a substance producing new substances with new properties.

CLOUD POINT – The temperature of a liquid (fuel or lubricant) at which a haze or a cloud first appears in a sample of oil, when cooled in a prescribed manner.

COAL – A firm, brittle, sedimentary, combustible rock derived from vegetable debries which have undergone a complex series of chemical and physical changes during the course of many million years.

COAL GAS – A fuel formed by the distillation of coal, usually in a retort or a coke oven.

COEFFICIENT OF HAZE – A measurement of visibility interference in the atmosphere.

COKE – A fused cellular porous structure that remains after the free moisture and the major portion of the volatile matter have been distilled from coal.

CAKING COALS – Coals that become soft under the usual furnace temperatures and merge into undesirable masses of coke. The coal that becomes soft, melts and solidifies into a more or less solid mass which further hardens on heating out of contact with air.

COLLOIDAL FUEL – A mixture of fuel oil and powdered coal.

COMBUSTION – Process involved during quick burning. Release of chemical energy into heat energy occurs during combustion.

COMBUSTION EFFICIENCY – is the ratio of the (heat) energy liberated to that which could be liberated under ideal conditions. Quantity of CO_2 and H_2O in the exhaust indicate energy liberated, whereas the quantity of H_2, CO and CH_4 indicate unliberated energy.

CRACKING – The process of breaking of heavy molecules into lighter hydrocarbons.

COMPRESSION IGNITION – Ignition of fuel through the heat of compression as in a diesel engine.

COMPRESSED NATURAL GAS – usually assumes the from of compressed methane, and is suitable for obtaining ultra low emissions from combustion engines. Soot production is virtually zero.

COMPOUND – A combination of two or more ingredients mixed together.

CONSTANT PRESSURE COMBUSTION – Combustion which occurs without a change in pressure. In an engine, this is obtained by the slower rate of burning than with constant volume combustion.

CRUDE OIL – Petroleum as it comes from the oil well (raw or refined state). It forms the basis of gasoline, engine oil, diesel oil, kerosene etc.

DETONATION – An uncontrolled instantaneous second explosion in a spark ignition engine, after the spark occurs, with excessively rapid burning of a portion of the compressed air fuel mixture (end charge almost exploding) resulting in a spark knock, or pinging noise.

DIESEL INDEX – A rating of fuel according to its ignition qualities. The higher the diesel index number, the better the ignition quality of the fuel.

DIMETHYLETHER – is a synthetic product with a high cetane number, producing little soot and reduced nitrogen oxide when combusted in diesel engines.

DISTILLATION – Heating a liquid, and then catching and condensing the vapours given off by the heating process.

DRAFT – The differential pressure in a furnace to ensure the flow of gases out of the furnace and flow of air into the furnace.

DUST – Fine grain particles light enough to be suspended in air.

ECOSPHERE – The layer of earth and troposphere inhabited by or suitable for existence of living organisms.

EFFLUENT – Waste material discharged into the environment, treated or untreated.

EGR SYSTEM – Exhaust gas recirculation system. It sends part of the exhaust gas back through the engine by way of the carburettor or intake manifold, which reduces the amount of NO_x that is formed by an engine.

ELECTOSTATIC PRECIPITATOR – An air pollution control device in which solid or liquid particulates in a gas stream are charged as they pass through an electric field and precipitated on a collection surface.

ELUTRIATION – A process of separating lighter particles from heavier particles by washing solid waste with a slowly moving upward stream of fluid that carried the lighter particles with it.

EMISSION CONTROLS – A term applied to any device or modification added onto, or designed into a motor vehicle for the purpose of controlling a source of air pollution emission.

ETHANOL – Ethyl alcohol produced by east fermentation of a variety of carbohydrates such as saccharin (sugar canes, sugar beets, molasses and fruit juices), starch (cereals and potatoes) or cellulose (wood waste, sulphite liquor).

ETHYL GASOLINE – Gasoline to which ethyl fluid has been added to improve its resistance to knocking. Slows down burning rate and thereby creates a smooth pressure curve that will allow the gasoline to be used in high compression engines.

EVAPORATIVE EMISSION CONTROL SYSTEM – A system which prevents the escape of gasoline vapours from the fuel tank or carburettor float bowl to the atmosphere while the engine is off. The vapours are stored in a canister, or in the crankcase until the engine is started.

EXCESS AIR – Air present in the cylinder over and above that which is theoretically necessary to burn the fuel.

EXHAUST GAS – The products of combustion coming out from an internal combustion engine.

EXHAUST GAS ANALYZER – A device for sampling the exhaust gas from an engine to determine the amounts of pollutants in the exhaust gas. This determines combustion efficiency.

FLAME DETECTOR – A device that monitors the flame in a furnace that is burning oil, gas, or pulverized coal fuel. Failure of the flame results in a signal and the actuation of various protective controls on the fuel feed to prevent an explosion.

FLAME SAFEGUARD SYSTEM – An arrangement of flame detection system, interlocks and relays, which will sense the presence of a proper flame in a furnace and cause fuel to be shut off to the furnace if a hazardous (improper flame or combustion) condition develops.

FLASH POINT – It is the temperature at which the quantities of vapour which a combustible fuel emits into the atmosphere are sufficient to allow a spark to ignite the vapour air mixture above the fluid.

FLUE DUST – Solid particles (smaller than 100 microns) carried in the products of combustion.

FLUE GAS ANALYZER – Device which measures the percentages of volume of carbondioxide, carbon monoxide and oxygen in the flue gas of a boiler.

FLY ASH – Combustion ash so fine that is carried up and into the atmosphere by the movement of the flue gases. It can become neighbourhood nuisance by settling on surfaces in the area after it looses its velocity.

FOG – Suspended liquid particles formed by condensation of vapour.

FORCED DRAFT FAN – The fan that pushes or forces air into the furnace, usually at a pressure higher than atmospheric pressure.

FRACTIONAL DISTILLATION – A process of separation on the molecular basis or on the basis of the boiling point of various fractions.

FREE LIQUIDS – Liquids which readily separate from the solid portion of a waste under ambient temperature and pressure.

FUEL – The substance that is burned to produce heat and create motion in an engine or heat substances. Any combustible substance.

FUEL CALORIEMETER – A meter (also called oxygen bomb) to determine the heating value of 1 kg of fuel by burning a sample of the fuel under controlled conditions.

FUGITIVE DUST – Particulate matter composed of soil which is uncontaminated by pollutants resulting from industrial activity. Fugitive dust may include emissions from haul roads, wind erosion of exposed soil surfaces and soil storage piles, and other activities in which soil is either removed, stored, transported, and redistributed, also solid air borne particulate matter emitted from any source other than through a stack.

FUME – Any kind of noxious vapour arising from a process of combustion or chemical reactions. Includes smoke, odorous materials, metallic dust.

FUME AFTERBURNERS – Units designed to consume combustible fumes by means of a direct fired combustion chamber through which the fumes must pass on their way to the stack and the atmosphere.

FURNACE EXPLOSION – The ignition and almost instantaneous explosion of highly inflammable gas, vapour or dust accumulated in a boiler setting.

GAS – A state of matter, neither solid nor liquid which has neither definite shape nor definite volume. Air is a mixture of several gases.

GASOLINE – A liquid blend of hydrocarbons, obtained from petroleum crude oil, used as the fuel for most automobile SI engines.

GRINDABILITY – A descriptive term of a characteristic of coal that is important to pulverized coal systems.

HEATING VALUE OF A FUEL – The heat liberated by the complete and rapid burning of a fuel per unit weight or volume of the fuel. Also called calorific value of the fuel.

HIGHER HEAT VALUE – A standard recommended by the ASME, the higher heat value of a fuel includes the heat value of the hydrogen in the fuel. The heating value indicated by a fuel caloriemeter.

HIGH TEST GASOLINE – A term referring to the octane rating of a fuel. A high test fuel has a high octane rating.

HOGGED FUEL – Wood that has been chipped and shredded, usually by a machine called a "hog".

HYDROCARBON – A compound made of elements of hydrogen and carbon atoms. Gasoline, diesel oil are blends of different hydrocarbons refined from crude oil.

IGNITION TEMPERATURE – The temperature at which the heat that is generated by the reaction between air and fuel vapour, is faster than that is lost to the surroundings, and combustion thus becomes self propelling. Below this point, the gas air mixture will not burn freely.

IGNITION QUALITY OF DIESELS – is indicated by cetane number. It is the percentage of cetane by volume, in a mixture of cetane ($C_{16}H_{34}$) and alpha methyl naphthalene which will exhibit the same ignition characteristic of the fuel under test when tested in a standard engine, under a set of standard test conditions.

INCINERATION – The controlled process in which the combustible solid, liquid or gaseous wastes are burned and changed into non-combustible gases.

INCINERATOR – Any furnace used in the process of burning waste for the primary purpose of reducing the volume of the waste by removing combustible matter.

INDUCED DRAFT FAN – The fan that draws the gases out of the furnace by creating a partial vacuum on the suction side of the fan.

INFRARED GAS ANALYZER – A non-dispersive infrared gas analyzer used to measure very small quantities of the pollutants contained in the exhaust gas.

KEROSENE – This petroleum product is a liquid fuel having an average latent heat of vapourization of 105-110 BTU lb and the specific heat of 0.50, sometimes called COAL OIL.

KNOCK (engine) – In an engine, a rapping or hammering noise resulting from excessively rapid burning of the compressed air fuel charge.

LIGNITE – A coal of high moisture content and low calorific value, generally less than 8300 BTU/lb. May require predrying before being used as a fuel.

LIQUID ASH REMOVAL SYSTEM – An arrangement of piping by which molten ash is removed continuously or intermittently, as desired, from the bottom of a furnace. The operating medium is usually compressed air with pneumatic controls.

LIQUEFIED PETROLEUM GAS (LPG) – A gas fuel that is stored in liquid form and is converted into gas as it leaves the storage tank by a pressure regulator that steps down the storage pressure on the liquid at the tank outlet and thereby permits the liquid to assume its normal gaseous state at the existing temperature and reduced pressure.

LONG FLAME BURNER – An oil or gas burner in which the mixture of fuel and air is delayed long enough to produce a long flame from the burner nozzle. Can be a source of trouble if the flame impinges on either refractory or tube surfaces.

LOWER HEATING VALUE – Net heat liberated per kg of fuel after the heat necessary to vapourize and superheat the steam formed from the hydrogen (and from the fuel) has been liberated.

LOW LEAD FUEL – Gasoline which is low in tetraethyl lead (approximately 0.5 gm per gallon).

MECHANICAL DRAFT – Draft produced artificially by means of forced or induced draft fans.

METHANOL – Methyl alcohol produced from coal by its liquifaction, by pyrolysis, or by its reaction with high pressure hydrogen.

NATURAL DRAFT – Draft produced by a chimney, by a column of hot gases existing inside the chimney.

NATURAL GAS – Gas obtained from petroleum mines.

NO_x – Oxides of nitrogen, a byproduct of combustion within the combustion chamber at high temperature and under heavy load. A basic air pollutant.

NO_x CONTROL – Any type of device, or system, used to reduce the amount of NO_x produced by an engine.

NO LEAD FUEL – Gasoline to which there has been no intentional addition of lead compounds.

OCTANE RATING – The measure of antiknock property of gasoline. The higher the octane rating, (OCTANE NUMBER), the more resistant the gasoline is to knocking or detonation and better the quality: Higher compression engines require higher octane gas.

OCTANE NUMBER OF A FUEL – is the percentage by volume of iso-octane in a mixture of iso-octane ($C_8 H_{18}$) and *n*-heptane ($C_7 H_{16}$) which will exhibit the same antiknock characteristic of the fuel under test when tested in a standard CFR variable compression ratio engine, under a set of standard test conditions.

OIL BURNER – Any device wherein oil fuel is vapourized or so called atomized and mixed with air in proper proportion for combustion.

PARTICULATES – Small particles of lead and other substances occurring as solid matter in the exhaust gas.

PEAT – A substance of vegetable origin always found more or less saturated with water in swamps and bogs.

PETROLEUM – Crude oil as it comes out of the ground, which consists of 83-87 per cent carbon, and 10-14 per cent hydrogen, plus traces of oxygen, nitrogen and sulphur. From the crude oil, gasoline, diesel, lubricating oil and other products are refined.

PHOTOCHEMICAL SMOG – The result of sunlight reacting with hydrocarbons and nitrogen oxides in the atmosphere.

PING – The sound resulting from sudden auto ignition of the airfuel charge in a SI engine combustion chamber. Characteristic sound of detonation.

POLLUTANTS – Any gas or substance in the exhaust gases from the engine or that evaporates from the fuel tank or carburettor. These gases or substances add to the pollution of our atmosphere.

POLLUTION – The presence of matter or energy whose nature, location or quantity produces undesirable environmental effects.

POSITIVE CRANKCASE VENTILATION – PCV system-A crankcase ventilating system which produces the circulation of air through the crankcase, thus clearing it of water vapour, unburned hydrocarbons, and blow by, the air passes into the intake system of the engine and hence into the combustion chambers where they are burnt.

POT TYPE BURNER – It is a hot plate burner in which the fuel oil drops into a hot plate and vapourizes.

POUR POINT OF FUEL – is the temperature at which crystals begin to appear and the fuel flow will be interrupted, as the fuel is being cooled.

PREI-GNITION – Ignition of air fuel mixture in the SI engine cylinder (by any means) before the (ignition) spark occurs at the spark plug terminals.

PRIMARY AIR – The air mixed with the fuel at or in the burner. It ensures instant combustion as the fuel enters the furnace.

PRIMARY POLLUTANT – A pollutant emitted directly from a polluting source.

PROPANE – A type of LPG that is liquid below –42°C at atmospheric pressure.

PULSATION – A panting of the flames in a furnace, indicating cyclic and rapid changes in the pressure in the furnace.

PURGE – The evacuation of air or any other designated gas from the duct line, pipe line, container or furnace. Purging may be done in some instances simply by the use of a fan or blower, in others by driving out the air or gas by means of an inert gas, such as nitrogen, under high pressure.

RADIOACTIVE – Substances that emit rays either naturally or as a result of scientific manipulation.

REFUSE – A term generally used for all solid waste materials.

RETORT – A trough or channel built into an underfeed stoker through which the stoker ram pushes green coal into the fire. The coal enters the fire from below, hence the name "underfeed".

ROTARY BURNER – One in which the oil entering at the center of a rotary cup is whirled around very rapidly until the oil is thrown away from the cup. By centrifugal force it mixes with air and ignites.

SAFETY CONTROLS – Devices that guard against (1) overpressure leading to explosions from the water side or steam side, (2) overheating of metal parts, possibly also leading to explosion in a fired boiler, (3) fire side explosions (furnace explosions) due to uncontrolled combustible mixtures on the firing side.

SECONDARY AIR – Air introduced into a furnace above and around the flames as may be necessary to promote combustion. This air is in addition to the primary air which enters either as a mixture with fuel or as blast underneath a stoker.

SHREDDER– A machine used to break up waste materials into smaller pieces by cutting or tearing.

SCRUBBER – A device that uses a liquid spray to remove aerosol and gaseous pollutants from an air stream.

SLACK – A coal of fine size, often screenings, maximum size is not likely to exceed 62.5 mm.

SLAG TAP FURNACE – A furnace for burning pulverized fuel in which the ash puddles in the bottom of the furnace in a molten state and is removed periodically or continuously, depending on the design of the system, while still in the molten condition.

SMOG – A term coined from smoke and fog. This is applied to the fog like layer that hangs over many areas under certain atmospheric conditions. Smog is compounded from smoke, moisture and numerous chemicals which are produced by combustion and from numerous natural and industrial processes.

SPECIFIC GRAVITY – A measure of the weight per unit volume of a liquid as compared with the weight of an equal volume of water.

SURFACE IGNITION – Ignition of air fuel mixture in the combustion chamber produced by hot metal surfaces or heated particles of carbon.

SYNTHETIC FUELS – Fuels such as ethanol and butanol derived from coal, and hydrogen from water.

TETRA ETHYL LEAD – A chemical put into engine fuel which increases octane rating, or reduces knock tendency. Also called ethyl and tel.

TORCH – Combustible material on a metal rod, such as oil soaked rags, used to light up oil and gas burners. The torch is extinguished by being plunged into a prepared receptacle.

TUYERES – Castings appearing as components of underfeed stokers and designed to admit air to the green coal moving through the retorts.

VAPOURIZATION – To change a liquid into a vapour, often by the addition of heat.

VAPOURIZING BURNER – A burner in which the fuel oil is vapourized by heating in a retort. It may be a mixing or non mixing type.

VAPOUR LOCK – A condition in the fuel system in which gasoline has vapourized and turned to bubbles in the fuel line or fuel pump, so that the fuel delivery to the carburettor of a SI engine is prevented or retarded.

VOLATILITY – refers to the ease with which a liquid vapourizes. A liquid which vapourizes at a relatively low temperature has a high volatility. This liquid is said to be highly volatile.

WIND BOX – A plenum from which air is supplied to a stoker or to gas or oil burners.

CHAPTER 17

STEAM BOILERS

AIR PREHEATER – A device that makes the final heat recovery from boiler flue gases and uses the same to preheat the incoming furnace air for its reaction with fuel.

ANTI-INCRUSTATOR – A substance used to prevent the formation of scale on the internal surfaces of steam boilers.

ANTIPRIMING PIPE – A pipe placed in the steam space of a boiler, so as to collect the steam while excluding entrained water.

AUTOMATIC FEED WATER REGULATOR – Device that regulates feedwater supply to the boiler according to load, and so does away with hand operation of valves on feed lines. It is controlled by temperature, its action depends upon expansion and contraction of some metal part.

AUTOMATIC INJECTOR – One that is self starting after its operation has been stopped by the interruption of its water supply.

BABCOCK and WILCOX BOILER – A water tube boiler consisting in its simplest form of a horizontal drum from which is suspended a pair of headers carrying between them an inclined bank of straight tubes.

BAG – A bulged out section of a portion of the shell, extending through the full thickness of the shell, caused by over heating and pressure.

BALANCED DRAUGHT – A system of air supply to a boiler furnace, in which one fan forces air through the grate, while a second, situated in the uptake, exhausts the flue gases. The pressure in the furnace is thus at atmospheric *i.e.*, is balanced.

BANKING LOSS – The fuel used in maintaining a floating bank or to maintain a dead bank and then raise the steam pressure to normal.

BANKING UP – Reducing the rate of combustion in a boiler furnace by covering the fire with slack or fine coal.

BENSON BOILER – A high pressure boiler of the once through type in which water is pumped through the successive elements of the heating surface, firing being by gas, oil, or pulverized coal.

BLISTER – A separation of the metal from the shell plate, caused by impurities rolled into the shell plate when formed.

BLOWDOWN OF SAFETY VALVE – The difference between the pressure at which the safety valve pops and that at which it closes.

BLOWING OFF – Act of letting out water and steam from a boiler to carry off accumulated mud and scale.

BLOW OFF VALVE – The valve which empties the boiler for cleaning, inspection, or repair. It blows out mud, scale, or sediment when the boiler is in operation and prevents excessive concentration of soluble impurities in the boiler. Also used for rapid lowering of boiler water level if it is too high.

BOILER – A closed pressure vessel in which a fluid is heated and converted to vapour for use external to itself, by the direct application of heat resulting from the combustion of fuel (solid, liquid or gaseous) or by the use of electricity or nuclear energy.

BOILER CAPACITY –The weight of steam, usually expressed in kg/hour, which a boiler can evaporate, when steaming at full load output.

BOILER COMPOSITION – Chemicals introduced into the boiler feed water to inhibit scale formation and corrosion, or to prevent priming or foaming.

BOILER CROWN –The upper rounded plates of the boiler of shell type.

BOILER EFFICIENCY – The ratio of heat supplied by a boiler in heating and evaporating the feed water to the heat supplied to the boiler in the fuel. It may vary from 60 to 90 per cent.

BOILER PATCH – A small piece of metal used to cover and strengthen a weak spot. A soft patch is a covering over a leak or defect which is fastened with bolts, as distinguished from a hard patch which is riveted.

BOILER PLATE – Mild steel plate, generally produced by the open hearth process, used mainly for the shells and drums of steam boilers.

BOILER PRESSURE – The pressure at which steam is generated in a boiler.

BOILER SETTING – The supporting structure on which a boiler rests, usually of brick for land boilers and steel for marine boilers.

BOILER TEST – (1) A hydraulic pressure test applied to check water tightness under pressure greater than the working pressure. (2) An efficiency test carried out to determine evaporative capacity and magnitude of losses.

BOILER TRIAL – An efficiency test of a steam boiler, in which the weight of feed water and of fuel burnt are measured and various sources of losses are assessed.

BOILER TUBES – Steel tubes forming part of the heating surface in a boiler. In water tube boilers, the hot gases surround the tubes. In locomotive and some marine boilers (fire tube boilers) the gases pass through the tubes.

BREECHING – The metal duct that carries the smoke and gases of combustion from a furnace to the stack or chimney for ultimate discharge to the atmosphere.

CARRY OVER – It is entrained moisture and associated solids passing from a boiler with the steam.

CAULKING – Upsetting or burring up of the edge of the plate or strap after riveting so as to make the edges press down tightly on the plate beneath and thus form a water and steam tight joint.

CHECK VALVE – A form of non return valve used to control the flow of water as in pump operation.

CHIMNEY – A tall, hollow cylindrical column built of steel, brick or concrete used to produce the required natural draft effect.

CHIMNEY EFFECT – The upward movement of warm air or gas, compared with ambient air or gas, due to the lesser density of the warm air or gas. Chimney effect may be a cause of uneven heating in buildings two or more stories high.

CLEANING THE FIRE – Operation of removing clinkers, etc., from the burning coal at regular intervals.

CLOSED HEATER – A type of heater in which the steam and feed water are separated by a metal surface.

CLYDE BOILER – A boiler similar to a scotch boiler, but instead of a water space at the back end of the combustion chamber, a removable back which is lined with some insulating material such as asbestos or fire tile, is existing.

COCK – A device for regulating the flow of fluids through a pipe.

COLLECTOR or DRY PIPE – A pipe placed inside a boiler at a high point and having small perforations throughout its length so as to take off steam at a multiplicity of points and thus avoid turbulence caused by taking off steam at only one point.

CORROSION – Chemical action which causes destruction of the surface of a metal by oxidation, rusting. It is an electrochemical attack.

DOUBLE TUBE BOILER – Boiler having an auxiliary tube placed inside each main tube. Gases flow from one end to the other end through the auxiliary tube and from that end to the other end in the reverse direction through the annular space in the main tube.

DOWN COMER – A large vertical tube or pipe for circulating water from the water space of the steam drum to water wall headers.

DRAFT – The difference of pressure producing air flow through the boiler furnace, flue and chimney.

DUPLEX PUMP – A combination of two pumps arranged side by side and so connected that movement of each operates the steam valve of the other.

ECONOMIZER – Feed water heater placed between the boiler and chimney (exhaust) stack, so as to absorb a portion of the heat in the gases, not absorbed by the boiler.

ELECTRIC BOILER – Unit in which the resistance of water between solid metal electrodes generates heat and thereby produces steam.

EROSION – It is a mechanical action causing wear by abrasion.

EXTERNALLY FIRED BOILER – One in which the furnace is outside the boiler shell, the furnace walls being usually of fire brick. Example, the familiar horizontal return tubular boiler.

FEED WATER – The water (chemically treated water) supplied to a boiler to replace that evaporated as steam or blown off.

FEED WATER HEATER – An apparatus for raising the temperature of boiler feedwater by abstracting some of the heat from exhaust steam or from the hot gases of combustion.

FEED WATER REGULATOR – An automatic device which controls the amount of feed water admitted to the boiler so as to maintain a constant water level in the boiler drum.

FIREBOX BOILER – Boiler having the fire within a firebox, although external to the shell, is rigidly connected to it.

FIRE CRACKS – Cracks caused by radiant heat, usually around circumferential riveted seams of thick plates.

FIRE LINE – The highest point of the heating surface in most common types of boilers.

FIRE TUBE – Tube in which the products of combustion pass through and water surrounds the tube.

FLASH BOILER – A boiler consisting of a series of coils of steel tubing, water is supplied by a pump to the top coil, from where it circulates through the other coils, becoming heated in its descent and issuing from the lower coil, as highly superheated steam.

FLUID VAPOURIZER GENERATOR – A closed vessel in which a heat transfer medium, other than water, is vapourized under pressure by the application of heat.

FOAMING – It is severe priming or agitation of the water level due to dirty or impure water. Small, stable, non-coalescing bubbles are formed through the boiler water.

FOULING – A condition of the flue gas passages in a boiler or furnace that adversely affects the transfer of heat, usually in the form of soot or scale.

FURNACE – That part of the boiler designed for burning the fuel.

FUSIBLE PLUG – A safety device which acts in case of dangerously low water. It consists of an alloy of tin, lead and bismuth and a covering of brass or cast iron. The plug melts during dangerous water levels and permits steam to rush into the furnace and put out fire.

GALLOWAY TUBES – Transverse tubes placed in a flue tube and attached to the openings in the side of the flue to increase the heating surface.

GAUGE COCK – A device for determining the water level in the boiler.

GIRDER STAY – A cast steel or built up girder with its ends resting on the side or end sheets of the fire box or combustion chamber, and supporting the flat crown sheet or top sheet of the combustion chamber by means of bolts.

GOOSE NECK – A short length of pipe having one complete turn to which the steam gauge is attached.

GRATE SURFACE – The area of the grate upon which fire rests in a coal or wood fired boiler.

GROOVING – Surface cracking of boiler plates. Expansion and contraction of parts too rigidly connected cause grooving.

GROSS FEED WATER – The net feed water plus the quantity of water provided for that blown out.

GUSSET STAY – Pieces of plate iron secured to the boiler front or back near the top or bottom by means of angle irons.

HEATING SURFACE IN A BOILER – Fire side area in a boiler exposed to the products of combustion. The area consists of the surface area of tubes, fire boxes, shells, tube sheets and projected area of headers.

HORIZONTAL RETURN TUBULAR BOILER – One so arranged that the products of combustion after passing along the length of the shell, return in an opposite direction through the tubes, before passing up the stack.

HOT WATER HEATING BOILER – A hot water heating boiler is used for space water heating, with the water returned to the boiler.

HOT WATER SUPPLY BOILER – A boiler which supplies hot water to be used externally to itself for washing, cleaning etc.

INCRUSTATION – A coating over, the coating, being commonly known as scale.

INDUSTRIAL BOILER – A stationary water tube steam generator, in which some of the steam is produced in a convective tube bank.

INJECTOR – An instrument for forcing water into a boiler against the boiler pressure by means of a steam jet.

INTERMITTENT BLOW DOWN – Blow down that is taken from the bottom of the mud-drum, water wall headers, or lowest point in the circulation system, at regular intervals.

INTERNALLY FIRED BOILER – Boiler in which the furnace is within the shell, being surrounded by water.

INTERNAL TREATMENT – Treating water in the boiler while evaporation is taking place by chemically adjusting or balancing the boiler water to prevent scale formation, corrosion, steam contamination and embrittlement.

JAW STAY – A round bar having jaws forged at one end and a flat plate at the other inclined at the proper angle for riveting to the boiler shell.

LIGAMENT – The metal between tube holes in boiler practise. Section of the metal not cut away between two adjacent tube holes.

LOCOMOTIVE BOILER – A specially designed boiler, specifically meant for self propelled traction vehicles on rails.

LOG SHEET – A forced reminder to check certain components of a boiler to prevent trouble from developing later and to note if proper operation is taking place. A data sheet.

LOW WATER CUT OFF – A device that shuts down the boiler immediately if the water drops to a dangerously low level.

MAKE UP WATER – Additional water to be put into the boiler periodically to make up for the loss due to leakage or exhausting of steam without condensation.

MARINE BOILER – A low head type special design boiler meant for ocean cargo and passenger ships with an inherent fast steaming capacity.

MECHANICAL DRAFT – The draft artificially produced by mechanical devices such as fans and in some units by steam jets.

MECHANICAL STOKER – A device constructed to automatically feed fuel to a furnace. Its use results in more efficient combustion owing to constant instead of intermittent firing.

MINIATURE HIGH PRESSURE BOILER – A boiler which does not exceed the following limits: (1) 16" inside diameter of shell. (2) 5 cuft. gross volume exclusive of casing and insulation (3) 100 psi gauge. If it exceeds any of these limits, it is called a power boiler.

MISSISSIPPI COCK – A cock in which steam pressure keeps the cock closed and a push button is provided to open the cock.

NATURAL DRAFT – The draft caused by the difference in weight between the column of hot gas inside the chimney and a column of cool outside air of the same height and cross-section.

NET FEED WATER – The quantity of water necessary to supply a stated evaporation in a given interval of time.

NON-SECTIONAL BOILER – A boiler in which the tubes are divided into groups, each group communicating with a header at each end, making independent units.

ONCE THROUGH BOILER – A boiler or steam generator which receives feed water at one end of continuous tubes and discharges steam at the other end.

OPEN HEATER – An open chamber in which the exhaust steam and water to be heated are brought into intimate contact by spraying the water through the steam, both the water and condensate going to the boiler.

PACKAGED BOILER – A completely factory assembled boiler either water tube or fire tube, including boiler firing apparatus, controls and boiler safety devices.

PALM STAY – A round rod having forged on one end a plate or palm.

pH VALUE – It is a number between 0 and 14 indicating the degree of acidity or alkalinity.

PITTING – A form of corrosion resulting in a series of minute holes or pits eaten into the surface of the metal to a depth of sometimes 6 mm.

POP SAFETY VALVE – A valve so constructed that it opens very suddenly like a cork popping out of a champagne or sparkling burgandy bottle and remains open until the pressure is reduced a predetermined amount.

PORCUPINE BOILER – A boiler having a vertical drum into which are screwed a multiplicity of horizontal radial short tubes.

PORTABLE BOILER – A boiler mounted on a truck, barge, a small river boat, or any other such mobile apparatus.

POSITIVE INJECTOR – One with a hand operated overflow valve, which permits operation at high pressure by stopping the drizzle from the overflow.

POWER BOILER – A power boiler is a steam or vapour boiler operating above 15 psig and exceeding the miniature size boiler.

PRIMING – It lifts the water level and delivers steam containing spray of water. It is usually caused by forcing a boiler too hard or by a too high water level or a combination of both these causes.

RADIAL STAYS – Long threaded rods used in locomotive boilers, screwed through both the firebox crown sheet and wrapper sheet and the ends riveted.

REDUCING VALVE – An automatic throttle valve for use where low pressure steam for heating or process is taken from high pressure mains.

REGENERATIVE AIR HEATER – A type of air heater containing a rotor which comes in contact alternately with hot gases and air thereby transfers heat from hot gases to the air.

REHEATER – A superheater that heats steam which is let out from a high pressure turbine and after heating sends the steam to a low pressure turbine thereby increases the plant efficiency.

RETARDERS – Also called Spinners-These are helical strips or ribbons of metal centered in horizontal or vertical fire tubes of a fire tube boiler for increasing the wiping effect on the inner surfaces of the tubes, by the flue gases on their way to the chimney stack. The tubes cannot be cleaned by scraping or brushing until these strips are removed. Retarders also increase the boiler frictional resistance to the flue gases.

RINGLEMANN CHART – The chart used for comparing the smokes density when no instrument is available.

RIVETED JOINT EFFICIENCY – Ratio of the strength of a unit section of the joint to the same unit length of solid plate. Unit length usually taken is the pitch of the rivets (distance from centre to centre) in the row having the greatest pitch.

RIVETED STAYS – Stays in which the threaded ends are riveted instead of having a nut at each end.

SAFETY VALVE – A circular valve connecting the steam space of a boiler and loaded to such an extent that when the pressure of steam exceeds a certain point, the valve is lifted from its seat and allows the steam to escape. The valve is loaded either by weight or by a spring. The release of steam saves the boiler from explosion.

SCALE – Incrustation within a vessel caused by the mineral substances from the water. Scale is a result of the chemical effect of the heat and concentration. A hard coating, chiefly calcium sulphate. If the scale is excessive, it leads to overheating of the metal and ultimate failure.

SCOTCH BOILER – A horizontal boiler in which the combustion chamber at the end of the boiler shell is entirely surrounded by water.

SCUM SCOOP – Apparatus for blowing out water from the surface to remove fine particles of scale forming foreign matter.

SEPARATOR – The device that removes as much moisture as possible from steam after it leaves the boiler.

SINGLE TUBE BOILER – Boiler made up of plain tubes, and the gases flow from one end to the other end of the tubes.

SOCKET STAY – Also called stay bolt consisting of a rod and socket.

STATIONARY BOILER – A boiler which is installed permanently on a land installation.

STAY BOLT – Short stay bars to support flat surfaces that are only a short distance apart, such as inner and outer sheets of water legs in a locomotive boiler. Generally have screw threads cut at one end, and sometimes at both ends, to receive a nut.

STAY ROD or THROUGH STAY – A plain rod 1 1/4" to 2 1/2" in diameter having nuts and washers for fastening the ends to the plates.

STAY TUBE – A thick tube with threads on the ends, one end being larger than the other so that the tube may be slipped through the large hole.

STEEL ANGLE STAY – Two lengths of steel angle riveted together forming a T shape piece and riveted to the plate.

STEAM – Water in a semigaseous condition. It is a vapour than a gas, that is a substance between the purely liquid and gaseous states. Wet steam.

STEAM BOILER – A closed vessel in which steam or other vapour is generated continuously for use external to itself by the direct application of heat resulting from the combustion of fuel (solid, liquid, or gaseous) or by the use of electricity or nuclear energy.

STEAM GAUGE – A device for indicating gauge pressure as distinguished from absolute pressure.

STEAM LOOP – An ingeneous thermal pump consisting of an arrangement of piping wherein condensate is returned to the boiler.

STEAM SEPARATOR – An apparatus for separating out moisture that may be carried in suspension by steam flowing in pipelines, and for preventing this moisture from reaching and perhaps damaging engines, pumps, or other machinery that may be driven by the steam.

STEAM SPACE – The space above the water level in a boiler where steam gets collected until it is drawn off through the steam main.

STEAM TRAP – An automatic device which allows the passage of water but prevents the passage of steam. It is used to drain pipes of condensate.

STOP VALVE – A non-return valve having a hand wheel and screw stem which acts only to close the valve.

SUBMERGED TUBE BOILER – A vertical boiler having tubes extending from the lower tube sheet to an upper submerged tube sheet.

SUPER CRITICAL BOILER – A boiler that operates above the super critical pressure of 3206. 2 psi and 705.4°F saturation temperature.

THERMAL LIQUID HEATER – A closed vessel in which a heat transfer medium other than water is heated without vapourization, and the heated fluid gives up its heat and does useful work outside the closed vessel.

THROUGH TUBE BOILER – A vertical shell boiler having tubes extending from the lower tube sheet to the full length of the shell.

TRAVELLING GRATE or CHAIN GRATE – A type of overfeed stoker consisting of an endless grate composed of short sections of bars passing over sprockets at the front and rear of the furnace.

TUBE SHEET – A sheet of a water tube boiler where tubes are inserted, either in a flat sheet or a drum.

UNDERFEED STOKER – One in which the fuel is fed upward from underneath.

WASTE HEAT BOILER – A boiler which uses byproduct heat such as from a blast furnace in a steel mill, exhaust from a gas turbine, or by products from a manufacturing process. Waste heat is passed over the heat exchanger surfaces to produce steam or hot water for conventional use.

WATER ANALYSIS – Analyzing a water sample *i.e.*, process of finding out how much of the various impurities and other chemical substances are present in the water. The results are usually expressed in parts per million (ppm).

WATER COLUMN – A boiler fixture consisting of a cylindrical piece to which are attached the water gauge and gauge cocks, thus combining the two into one unit. The top and bottom have outlets which connect it with the boiler below and above the water level.

WATER GAUGE – A device used to indicate the height of water within a boiler.

WATER GRATE – A series of pipes connected close together in parallel to a header at one end and to upflow elements at the other.

WATER LINE – The level at which water stands in the boiler.

WATER TUBE – One which is surrounded by the products of combustion, the water being inside the tube.

WATER TUBE BOILER – A boiler which contains one or more relatively small drums with a multiplicity of tubes in which water steam circulate and hot gases surround the tubes.

WINDOW PATCH – A patch used to seal a hole cut in a water tube to provide access for welding the backside of a circumferential joint, or to replace a small, sharp bag.

CHAPTER 18

STEAM ENGINES AND STEAM TURBINES

ACCELERATED FLOW – Type of flow that takes place in nozzles. The flow accelerates and pressure reduces.

ADMISSION – The point in the working cycles of a steam or IC engine at which the inlet valve allows entry of the working fluid into the cylinder.

AIR PUMP – A reciprocating or centrifugal pump used to remove air, and sometimes the condensate, from the condenser of a steam plant.

ANGLE OF ADVANCE – The angle in excess of 90 degree by which the eccentric throw of a steam engine valve gear is in advance of the crank.

ALLAN VALVE – A steam engine slide valve, in which a supplementary passage increases the steam supply to the port during admission to reduce wire drawing.

ATMOSPHERIC ENGINE – An early form of steam engine in which a partial vacuum created by steam condensation allowed atmospheric pressure to drive down the piston.

AXIAL DISCHARGE TURBINE – A steam turbine in which the absolute velocity of steam flow at exit is a minimum *i.e.*, the steam discharges in the axial direction.

AXIAL FLOW TURBINE – Steam turbine in which the general direction of steam flow has been roughly parallel to the turbine axis.

BACK PRESSURE TURBINE – A steam turbine from which the whole of the exhaust steam, at a suitable pressure, is taken for heating purposes.

BAROMETRIC CONDENSER – A high level jet condenser.

BASTERED CONDENSER – It is an atmospheric keel condenser, which are sometimes fitted to canal boats or other sea vessels.

BINARY VAPOUR ENGINE – A heat engine using two separate working fluids, generally mercury vapour and steam, for the high and low temperature portions of the cycle respectively, thus enabling a large temperature range to be used, with improved thermal efficiency.

BLADE – Part attached to the rotating element of the machine or rotor, in which the stream of steam particles has its direction and hence its momentum changed. Also called DEFLECTOR.

BLADE VELOCITY COEFFICIENT – The ratio of the relative velocity of steam at outlet to the relative velocity at inlet of the blade.

BLANK FLANGE – A disc, or solid flange, used to blank off the end of a pipe.

BLAST PIPE – The exhaust steam pipe in the smoke box of a locomotive, which terminates in a nozzle to provide draft by entraining the flue gases in the steam jet and exhausting them through the chimney.

BLEEDING – A method of improving the thermal efficiency of steam plant by withdrawing a small part of the steam from the higher pressure stages of a turbine to heat the boiler feed water.

BLEEDER TURBINE – A steam turbine in which the steam is extracted at one or more intermediate stages for industrial use, often at comparatively high pressure.

BYPASS GOVERNING – Governing arrangement in which part of the steam that enters the turbine is bypassed depending upon the extent of load reduction.

CARRY-OVER LOSS – Loss of kinetic energy at the exit of the turbine. Also called LEAVING LOSS.

CHOKED FLOW – When a nozzle operates with the maximum mass flow it is said to the choked.

COMPOUND TURBINE – A multistage steam turbine in which the pressure energy of the steam is progressively transformed into kinetic energy in two or more stages with or without velocity compounding in each stage.

COMPOUND STEAM ENGINE – An engine which has two or more cylinders of successively increasing diameters, so arranged that the exhaust steam from the first cylinder (high pressure cylinder) is passed on to do work in the second cylinder (low pressure cylinder), and to a third cylinder in triple expansion engine, before being finally exhausted into a condenser.

CONDENSER – A vessel into which steam is exhausted and condensed instead of being rejected into the atmosphere after doing work in an engine cylinder or turbine. This is primarily for removing the back pressure upon an engine or turbine and thereby improve the plant efficiency.

CONDENSING CYCLE – A steam power plant cycle in which the exhaust steam is discharged into a condenser having a low back pressure, so that more energy can be extracted per unit weight of steam.

CONSTANT VELOCITY FLOW – Type of flow that takes place in parallel ducts.

COOLING POND – A shallow reservoir having a large surface area for removing heat from the cooling water used to condense steam in condensers.

COOLING TOWER – An apparatus designed to remove from the cooling water, used in a condenser, as much heat as can possibly be abstracted per unit space occupied by the apparatus.

COUNTERFLOW STEAM ENGINE – The engine in which the steam leaves the cylinder at the same end at which it entered.

CRITICAL PRESSURE OF NOZZLE – The pressure at which the velocity of the fluid equals the local sound velocity.

CRITICAL PRESSURE RATIO – Ratio of critical pressure of nozzle to the initial pressure.

CRITICAL SPEED OF A SHAFT – The speed at which the shaft displacement tends to be very large, and the shaft may become permanently bent.

CUSHION STEAM – The steam present in the cylinder during compression which occurs just after the exhausting of steam by the inward movement of the piston.

CUTOFF GOVERNING – Control of engine speed is accomplished by changing the volume of steam admitted to an engine cylinder as the load fluctuates. The points of steam cut off comes early in the stroke of the engine piston with light loads and later when they increase.

DECELERATED FLOW – Type of flow that takes place in the diffusers velocity decreases.

DEGREE OF REACTION – In an impulse reaction turbine, it is the ratio of the enthalpy drop that takes place in rotor blades to the total enthalpy drop that occurs in the stage.

DIAGRAM EFFICIENCY or BLADING EFFICIENCY – Ratio of the rate of doing work per kg of steam (diagram work) to the energy supplied to the rotor per kg of steam.

DIAGRAM FACTOR – Ratio of actual indicator diagram area to the hypothetical indicator diagram area.

DIAPHRAGMS – Partitions, which separate one wheel chamber from the next and in which nozzles are usually fitted in the case of pressure compounded impulse turbine.

DIFFUSER – The duct in which a fluid is decelerated causing a rise in pressure along the stream.

DISC FRICTION – Surface friction loss due to relative motion between the disc and steam particles when the disc is rotating in the viscous fluid *i.e.*, steam.

DOUBLE ACTING ENGINE – A steam engine in which a power cycle is produced in each end of the cylinder during one revolution of the crankshaft.

DOUBLE DECK SPRAY POND – The pond having spray nozzles arranged at different elevations.

D SLIDE VALVE – A sliding valve which alternately admits steam to and releases the steam from each end of the steam engine cylinder.

DRY AIR PUMP – A pump designed to handle air and gases only, which will give a higher vacuum than possible with a wet air pump.

ECCENTRICITY – The distance between the centre of the crankshaft and the centre of the eccentric and is equal to the effective crank arm of the eccentric.

ENGINE RELATIVE EFFICIENCY – Ratio of actual thermal efficiency to that of the rankine cycle for the same pressure and temperature conditions.

EVAPORATIVE CONDENSER – A type of surface condenser in which the steam is fed through the condenser tubes over which cooling water is sprayed and cooling is effected mainly by evaporation of the sprayed water.

EXHAUST LAP – The overlap of the release edge of the D slide valve over the release edge of the port when the valve is in the mid position which is the middle of its travel.

FULL ADMISSION TURBINE – A steam turbine in which steam is admitted over the entire circumference of the blade annulus. Reaction turbines are full admission turbines.

GLAND – Device provided to minimise the leakage of steam, or in the case of the low pressure end of the turbine the leakage of air through the clearance space which separates the rotor from the casing.

HEAT ENGINE – A device which transforms heat energy into mechanical energy and render the latter available for doing useful work. It is merely an energy transformer.

HELICAL FLOW TURBINE – A single pressure, multiple velocity stage machine designed for helical flow.

IMPULSE – The act of impelling or suddenly driving forward in the same direction as the applied force.

IMPULSE TURBINE – A steam turbine in which the steam is expanded causing pressure drop in nozzles only and the moving blades attached to the rotor merely deflect the steam through an angle. The pressure on the two sides of the blades remains constant.

IMPULSE REACTION TURBINE – A steam turbine in which the steam is expanded both in the fixed blade and the moving blade continuously as the steam passes over them. The pressure drops gradually and continuously over both moving and fixed blades. Often called as REACTION TURBINE.

INTERNAL TURBINE EFFICIENCY – Ratio of the total useful heat drop to the adiabatic heat drop corresponding to the whole pressure drop that occurs. This is the product of stage efficiency and reheat factor.

JET CONDENSER – A closed chamber within which exhaust steam comes in direct contact with a spray or jet of cold water and is condensed.

KEEL CONDENSER – A type of marine outboard single pass surface condenser attached to the side of a hull below the water line. It requires no circulating water pump.

LAP OF D SLIDE VALVE – It is that portion of the valve face which overlaps the ports when the valve is in its central or neutral position.

The distance overlapping on the outside is called OUTSIDE LAP or STEAM LAP, and the distance overlapping on the inside is called INSIDE LAP or EXHAUST LAP.

LAYBRINTH GLAND – A type of gland consisting of a series of intricate passages, laybrinths, which are designed to destroy the kinetic energy developed by the steam leaving through the small clearance space. In the clearance space, this energy is converted to thermal energy.

LEAD OF D SLIDE VALVE – The amount by which the steam port is opened by the valve when the piston in a steam engine cylinder is at the begining of its stroke.

LOW PRESSURE TURBINE – A turbine operating at approximately atmospheric pressure and expanding to condenser pressure.

MISSING QUANTITY – In a steam engine, this is the lack of volume due to the steam not remaining dry and saturated during expansion.

MIXED PRESSURE TURBINE – A turbine designed to work on two or more pressures.

NATURAL POND – A natural flow pond having no baffle walls or spray nozzle.

NET EFFICIENCY – Ratio of network delivered at shaft to the total energy supplied to the steam turbine/steam engine. In the case of steam turbine it is the product of nozzle efficiency, diagram efficiency and mechanical efficiency.

NOZZLE – A duct of smoothly varying cross-sectional area in which a steadily flowing fluid can be made to accelerate by a pressure drop along the duct.

NOZZLE CONTROL GOVERNING – Governing arrangement in which nozzles are grouped together in three or five or more groups and each group of nozzles is supplied with steam which is controlled by valves.

NOZZLE EFFICIENCY – Ratio of the actual enthalpy drop to the isentropic enthalpy drop between the same pressures.

NOZZLE THROAT – The section of the nozzle where the area is minimum.

OVER EXPANDING – A nozzle which operates with a back pressure above the designed value of pressure at the exit of the nozzle.

OVERALL THERMAL EFFICIENCY – Ratio of the useful heat drop to the heat supplied in a turbine.

PARTIAL ADMISSION TURBINE – An impulse turbine in which the nozzles occupy only a part of the circumference leading into blade annulus and therefore admit steam over part of the blade annulus. Impulse turbines are partial admission turbines.

PORT – The entrance at the valve seat to either a steam passage leading to the cylinder or an exhaust passage leading to the exhaust pipe.

PRESSURE COMPOUNDED IMPULSE TURBINE – Steam turbine in which the total pressure drop of steam is divided into stages in a number of rings of fixed nozzles and the velocity produced by a ring of nozzles is utilized almost fully in the immediately following row of moving blades attached to the rotor.

PRESSURE VELOCITY COMPOUNDED IMPULSE TURBINE – Steam turbine in which the total pressure drop of steam is divided into stages and the velocity thus obtained in each stage is also compounded ie utilized in two or more rings of moving blades.

RADIAL FLOW TURBINE – A steam turbine in which the blades are so arranged to cause the flow of steam radially inwards or outwards.

RANKINE EFFICIENCY OF TURBINE – Ratio of the adiabatic heat drop to the heat supplied.

REAMED NOZZLE – A round nozzle used primarily in the high pressure impulse stage steam turbines. They have lower efficiency and somewhat greater in length.

REHEAT CYCLE – A system in which steam is initially expanded through a turbine and then reheated before further expansion in the turbine.

REHEAT FACTOR – In a multistage steam turbine it is the ratio of the sum of the individual heat drops (cumulative drop) in the different stages to the direct or adiabatic drop in a single step for the whole pressure drop that occurs.

RELATIVE EFFICIENCY – Ratio of the overall thermal efficiency of the actual plant to the thermal efficiency of the ideal Rankine cycle.

RIM HORSE POWER – Power developed by steam in passing over the blade as obtained from the velocity diagrams.

SCOOP CONDENSER – A type of marine condenser with a flow of circulating water induced through an enclosed chamber by the movement of the vessel, rather than externally as with a keel condenser.

SIMPLE IMPULSE TURBINE – A steam turbine in which the expansion of the steam takes place in one set of nozzles only.

SIMPLE MULTISTAGE TURBINE – A steam turbine in which the pressure energy of the steam is progressively transformed into kinetic energy in two or more pressure stages, there being one wheel to each stage.

SINGLE WHEEL IMPULSE TURBINE – A steam turbine having a set of nozzles and one row of moving blades fixed to the rotor. The moving blades deflect the steam and therefore cause a change in momentum and consequently produce force *i.e.*, motive force.

SPRAY POND – A pond which is provided with spray apparatus so that the hot cooling water from the condenser is sprayed over the surface of the pond and cooled.

STAGE EFFICIENCY – Ratio of the rate of doing work per kg of steam (diagram work) to the energy supplied to the stage per kg of steam, in a steam turbine. Also called GROSS STAGE EFFICIENCY. This is product of nozzle efficiency and blade efficiency.

STEAM CONDENSER – Device in which the exhaust steam from an engine or a turbine is condensed and air and other non condensable gases are removed in a continuous process.

STEAM JET VACUUM PUMP – Pump system in which one or more steam jets moving at a high velocity (1000 m/s) entrain the saturated non condensable gases, and the mixture passes through an expanding tube in which the velocity is reduced with a resulting increase in pressure and finally the mixture is let out into the atmosphere. Also called STEAM JET AIR EJECTOR.

STEAM LAP – The distance by which the admission edge of the D slide valve overlaps the edge of the steam port when the valve is in the mid position, which is the middle of its travel.

STEAM RATE OF AN ENGINE – The weight of steam supplied to a steam engine per horse power hour.

STEAM TURBINE – A prime mover in which gradual changes in the momentum of a fluid are utilized to produce rotation of the movable member.

STEAM TURBINE EXTERNAL LOSSES – Losses which do not affect the steam conditions while it flows through the turbine, includes mechanical losses and losses due to the end leakages.

STEAM TURBINE INTERNAL LOSSES – Losses connected with the steam conditions while it flows through the turbine. These include losses in regulating valve, nozzles, moving blades, carry-over losses from one stage to another, rotor and guide blade clearance losses, loss due to steam wetness and exhaust losses.

SURFACE CONDENSER – A device for condensing steam, in which the steam and the cooling water do not come into contact with each other, but are separated by metal surfaces.

SURFACE CONDENSER SINGLE PASS TYPE – Type of surface condenser in which the cooling water flows in one direction only through all the tubes.

SURFACE CONDENSER TWO PASS TYPE – Type of surface condenser in which the cooling water flows in one direction through part of the tubes and returns through the remainder.

TAPERED BLADES – Blades tapered in width along the length of the blade so as to obtain an even more uniform centrifugal stress.

THERMAL EFFICIENCY OF RANKINE CYCLE – Ratio of the thermal equivalent of output to thermal equivalent of input of a unit working on Rankine cycle.

THERMOCOMPRESSOR or BOOSTER EJECTOR – Compression device designed to handle steam instead of other gases.

THROTTLE GOVERNING – Governing effected in a steam engine by varying the initial steam pressure within the engine cylinder. The governor is called THROTTLE GOVERNOR. Power output is varied by varying initial steam pressure.

TRANSONIC NOZZLE – A nozzle usually of the reamed type and this is used in small impulse steam turbines, or velocity compounded stages where large enthaply drops are required.

UNDER EXPANDING NOZZLE – A nozzle which operates with a back pressure below the designed value of pressure at the exit of the nozzle.

VACUUM BREAKER – An automatic device used to protect the main engine or turbine from flooding when a jet condenser is used.

VACUUM KEEL CONDENSER – A type of keel condenser having tubular condensing surface, a return pipe for the condensate, the end of which connects with a wet air pump.

VELOCITY COMPOUNDED IMPULSE TURBINE – Steam turbine in which the total heat drop takes place in the nozzles, but the velocity (energy) generated is utilized in two or more rings of blades attached to the rotor.

WILLANS LINE – The line which represents the total weight of steam necessary per hour plotted as ordinates against the corresponding engine loads (either as IHP, BHP or KW) as abscissa.

WIRE DRAWING – In a steam engine, the fall in pressure caused by steam flowing through the restricted (valve) Passage immediately before the point of cut off.

WORKING STEAM – The steam that actually flows through the blades, this is different from the total steam flow through the turbine.

CHAPTER 19

GAS TURBINES

ADIABATIC PROCESS – Thermodynamic process in which no heat is transferred to or from the system during the process. A reversible adiabatic process is called ISENTROPIC PROCESS.

AIRFOIL – A streamlined form bounded principally by two flattened curves and whose length and width are very large in comparison with thickness. The airfoil may be a symmetrical airfoil or a non-symmetrical airfoil.

AIRFOIL DRAG FORCE – Force acting on the airfoil in the direction of motion, represents frictional forces.

AIRFOIL LIFT FORCE – Force acting on the airfoil in the direction perpendicular to the direction of motion. Basic force causing the aeroplane to maintain its lift.

AIR RATE – Kilograms per second of airflow required per net horse power developed. Also called FLOWRATE.

ANGLE OF ATTACK – The angle of inclination of the non-symmetrical airfoil with the direction of the undisturbed flow.

ANGLE OF DEVIATION – The difference between the fluid angle at outlet and the blade angle at outlet. This may be positive or negative. Sometimes called DEVIATION.

ANGLE OF INCDIENCE – The difference between the fluid angle at inlet and the blade angle at inlet. This may be positive or negative. Sometimes called INCIDENCE.

ANNULAR TYPE COMBUSTION CHAMBER – The combustion chamber which is made up of four concentric surfaces surrounding the axis of the rotor, forming three chambers on either side of the rotor, the middle casing acts as a flame tube and the inner and outer casings act as air casings, with a series of burners at the front end.

ASPECT RATIO OF BLADE – Ratio of blade height to blade chord.

AXIAL FLOW COMPRESSOR – A type of compressor in which the fluid flow is almost parallel to the axis of the compressor and the flow is decelerating or diffusing and pressure rises are obtained by causing the fluid to pass through a number of expanding spaces with consequent reduction in velocity.

BLADES or BUCKETS – The parts that form the rotor flow passages and serve to change the direction, and hence the momentum, of the fluid received from the stationary nozzles.

BLADE SPEED RATIO – Ratio of mean blade speed to the absolute velocity of the fluid stream at the blade inlet.

BOUNDARY LAYER – A thin layer of fluid adhering to a surface, when the fluid flows along the surface, in which there is a steep velocity gradient due to viscous friction, the velocity dropping to zero at the boundary surface.

BRAYTON CYCLE – Basic cycle for gas turbines. The cycle in which air is compressed isentropically, heated at constant pressure and expanded isentropically thus delivers work until the low pressure is reached and then heat is rejected. Also called JOULE CYCLE.

CAN TYPE COMBUSTION CHAMBER – Combustion chamber in which the air leaving the compressor is split into several streams and each stream is supplied to a separate cylindrical combustion chamber.

CARRY OVER LOSS – Kinetic energy discarded in the exhaust. Axial exit of the fluid from the turbine blades reduces this loss.

CASING – Turbine enclosure to which the nozzles and guides are fixed. Also called a SHELL or CYLINDER.

CENTRIFUGAL COMPRESSOR – A type of compressor in which air is sucked into the impeller eye, whirled around at high speed by the vanes on the impeller disc and flung out by centrifugal force.

CLOSED CYCLE TURBINE – Turbine in which the working fluid does not come in contact with the atmospheric air and the heat to the working fluid is provided in the heater by burning the fuel externally.

COMBINATION PLANT – A gas turbine plant that utilizes reheat, intercooling and regeneration.

COMBUSTION CHAMBER – The unit in which the chemical combination of oxygen in the air supplied by the compressor takes place with the carbon and hydrogen components of the fuel in such a manner that a steady stream of the gases at uniform temperature is produced and delivered to the turbine.

COMBUSTION EFFICIENCY – The ratio of the actual heat realised by the combustion of fuel to the ideal value *i.e.*, calorific value.

COMBUSTION INTENSITY – Ratio of the rate of heat supply by fuel to the product of volume of combustion chamber and inlet pressure in atmospheres.

COMPRESSOR EFFICIENCY – Ratio of work required for isentropic compression to the actual work input. Also called ISENTROPIC COMPRESSION EFFICIENCY.

COMPRESSOR MECHANICAL EFFICIENCY – Ratio of rotor horse power to shaft horse power supplied to the compressor.

CONSTANT PRESSURE TURBINE – Turbine in which the fuel is burnt at constant pressure. Combustion is a continuous process.

CONSTANT VOLUME TURBINE – Turbine in which the combustion takes place at constant volume. Also called EXPLOSION TYPE TURBINE.

CONVERGENT DIVERGENT DIFFUSER – A type of diffuser which can build up pressure when velocities are reduced from supersonic to subsonic values.

COOLING OF TURBINE BLADES – Turbine blades are cooled by water or air. This enables the temperature of the blade metal to be several hundred degrees lower than the gas temperature and permits employment of correspondingly higher turbine inlet temperatures, with the metals available at present, resulting in higher turbine efficiency.

COUNTERFLOW HEAT EXCHANGER – A heat exchanger in which compressed air and hot gases let out by the turbine, flow in opposite directions.

CROSS COMPOUNDED UNIT – The system in which the low pressure compressor is driven by the high pressure turbine and the high pressure compressor by the low pressure turbine.

CROSS FLOW HEAT EXCHANGER – A heat exchanger in which the compressed air and the hot gases let out by the turbine flow normal to one another.

CYCLE PRESSURE RATIO – Ratio of the pressure at inlet to the gas turbine to that at inlet to the compressor.

CYCLE WITH INTERCOOLED COMPRESSION – Gas turbine cycle in which the compression of the working fluid is cut off at some intermediate pressure and the fluid is cooled by passing it through a heat exchanger supplied with coolant from some external source before being compressed in the second compressor to the required pressure ratio.

DEFLECTION ANGLE – Total fluid turning angle *i.e.*, difference between the fluid angle at inlet and the fluid angle at outlet.

DEGREE OF REACTION (compressor) – The ratio of the static temperature rise in the rotor to that in the whole stage.

DEGREE OF REACTION (turbine) – Ratio of enthaply drop in rotor blades to enthaply drop in the stage.

DIAPHRAGM – The component fixed to the cylinder or casing and contains the nozzles and serves to confine the fluid flow to the nozzle passages.

DIFFUSER – Unit connected to the outlet of the centrifugal compressor, which provides a gradually increasing area to convert velocity energy into pressure energy.

DIFFUSION – Process in which the energy of a moving stream of fluid is transformed in such manner, then an increase in pressure occurs.

DISC or WHEEL – The component to which the moving blades are attached directly and it is keyed or shrunk on the shaft.

DISC FRICTION – When a disc rotates in free air, a certain amount of pumping action would take place, imparting motion to the surrounding air, and this relative motion between the disc and air causes friction, called disc friction.

DOUBLE SIDED IMPELLER – Impeller of a centrifugal compressor in which suction takes place from both sides. Here, two similar impellers are placed back to back.

DYNAMIC HEAD – The difference between the total head pressure and the static pressure.

EFFICIENCY OF IMPULSE BLADING – Ratio of the delivered power or energy, to the power or energy supplied in kinetic form to an impulse blade.

ERICSON CYCLE – The gas turbine cycle which incorporates multistage compression with intercooling, and multistage expansion with reheating.

FLAME STABILIZATION – Making the flame to be more or less stable at a particular location in the combustion chamber by a system whereby part of the high temperature products of combustion can be caused to recirculate in order to ignite fresh reactants.

FLOW COEFFICIENT – Ratio of axial components of absolute velocity of fluid stream to the blade velocity at that location.

FLOW LOSSES – Pressure loss due to friction and turbulence. This consists of combustion chamber loss, heat exchanger loss (air side), heat exchanger loss (gas side), intercooler loss (air side) and duct losses occurring between components and at intake and exhaust.

FLUID ANGLES – Angles at which the fluid enters and leaves a blade. These are seldom the same as the blade angles.

FOIL NOZZLE – A nozzle formed by curved airfoil sections or facsimiles of airfoils and is characterised by its high efficiency.

FREE VORTEX FLOW – In a compressor, the condition when the whirl velocity of a flowing fluid varies inversely as the radius.

FUEL RATIO – The weight of fuel used to heat unit weight of compressed air to the turbine inlet temperature.

FULL ADMISSION – Admission of gas stream over the full blade entry, and this becomes possible when the nozzles subtend the whole blade circumference (annulus area).

GAGING – Ratio of the net area of gas flow to the total free annular area in the blade ring.

GAS TURBINE – A rotary machine, which consists of a compressor, combustion chamber and a turbine. Air is compressed in the compressor, passed into the combustion chamber where fuel is burnt, products of combustion impinge over rings of turbine blades with high velocity and work is done.

GUIDE BLADES – Row of blades interposed between the blade rows comprising rotor passages, so as to reverse the direction of the fluid leaving the preceding moving blade row and make the general direction of the fluid entering all the moving blade rows to be similar.

HEAT EXCHANGER – Device which uses some of the heat in the turbine exhaust gas to preheat the air entering the combustion chamber. This reduces the fuel supply for a given required temperature increase. Also called REGENERATOR.

HEAT EXCHANGER THERMAL RATIO – Ratio of the actual heat picked up by the compressed air in the heat exchanger to the maximum possible heat that could be absorbed. Also called EFFECTIVENESS OF HEAT EXCHANGER.

INTER COOLING – Cooling of air in between the stages of compression so as to reduce the work of compression. The device used for cooling is intercooler.

INTERNAL EFFICIENCY OF A TURBINE – Ratio of work (power) delivered to the rotor of the turbine by the gaseous medium compared with the ideal energy available for work (power) from the medium in expanding through the same pressure range.

ISENTROPIC EFFICIENCY – Ratio of work to compress isentropically to the actual work to compress. Also the ratio of isentropic temperature rise to the actual temperature rise.

JET PROPULSION – The unit in which the gas turbine is designed to produce just sufficient power to drive the compressor and the exhaust gases from the turbine are then expanded to atmospheric pressure in a propelling nozzle to produce a high velocity jet.

LACING WIRES – Wires called lacing wires or lashing wires are used to keep long blades in alignment and to add stiffness.

LEAKAGE LOSSES – Energy loss due to leakage of the working fluid in turbines, between stages, past the shaft and around the balance piston.

MACH NUMBER-M – Ratio of the stream velocity to the local acoustic velocity.

MULTI SHAFT GAS TURBINE – Turbine unit in which two or more compressor combinations or turbines are carried on independent shafts. In each turbine compressor combination, the machines are coupled to each other in a series arrangement.

NET JET THRUST – That part of the thrust of a turbojet engine which is available for climb and acceleration.

NOZZLE – A flow passage specially shaped to produce kinetic energy at the expense of other forms of energy (available thermal energy).

NOZZLE EFFICIENCY – The ratio of the actual kinetic energy produced on discharge (or between any two points in a nozzle) to that obtainable by assuming an isentropic expansion in the nozzle.

ONE DIMENSIONAL FLOW – The fluid flow in which the variables are constant over any cross-section of the flow.

OPEN CYCLE TURBINE – Turbine in which the heat is transferred by direct combustion and after doing work in the turbine, the gases are exhausted into the atmosphere.

OVERALL EFFICIENCY OF JET PROPULSION – Product of the propulsion efficiency of the jet unit and the thermal efficiency of the jet unit.

OVERALL EFFICIENCY OF PROPELLER UNIT – Product of the propeller (propulsion) efficiency, thermal efficiency of the engine (power turbine) and the transmission efficiency from prime mover to propeller shaft.

OVERALL TURBINE EFFICIENCY – Ratio of the delivered shaft work (power) to the ideal energy available from the medium.

PARALLEL FLOW HEAT EXCHANGER – A heat exchanger in which both compressed air and gases let out by the turbine flow in the same longitudinal direction. Also called UNI DIRECTIONAL or COCURRENT HEAT EXCHANGER.

PARTIAL ADMISSION – Admission of gas stream over only a part of the blade circumference, since the nozzles are covering only a fraction of the inlet blade circumferential annulus. This is the case in impulse turbines.

PLANE SHOCK WAVE – Shock wave in which the variables of flow ego pressure, temperature and velocity are constant along the wave front.

POLYTROPIC EFFICIENCY – The isentropic efficiency of an elemental stage of the compression which is constant throughout the process. Also called the SMALL STAGE EFFICIENCY.

POSITIVE DISPLACEMENT COMPRESSOR – Type of compressor in which a fixed amount of working fluid is being positively contained during its passage through the machine.

POWER INPUT FACTOR – For a compressor this is the ratio of the actual work to the theoretical work of compression. Represents an increase in the work input the whole of which is absorbed in overcoming frictional loss and which is therefore degraded into heat energy.

POWER RATIO – Ratio of useful or net horse power of the cycle compared with the power developed by the turbine of the system. Also called WORK RATIO.

PRESSURE COMPOUNDED IMPULSE TURBINE – The turbine in which the pressure range available for expansion is broken into a series of steps or stages. Each stage consists of a nozzle or bank of nozzles (which increase the kinetic energy) followed by a row of turbine blades or buckets (which absorb the kinetic energy).

PRESSURE RATIO – Ratio of pressure of air at the end of compression to the pressure of air at the begining of compression.

PREWHIRL – Whirl velocity (tangential component of the absolute velocity at intake), imparted to the air that enters the centrifugal compressor impeller, by allowing the air to be drawn into the impeller eye over curved inlet guide vanes attached to the impeller casing.

PRIMARY AIR – Part of the air which flows through the core of the combustion chamber, in just sufficient quantity for combustion.

PRIMARY ZONE – Portion of the combustion chamber wherein about 15 to 20% of the air is introduced around the jet of fuel and the burning of this rich mixture provides the high temperature necessary to prepare the mixture for further reaction and burn it almost completely in a very short time.

PROPELLER JET ENGINE – The unit in which the gas turbine develops power in excess of that required to drive the compressor and employs this excess power to drive a propeller through reduction gearing and the leaving jet also contributes to the thrust power.

PROPULSION EFFICIENCY – Ratio of thrust power to the jet power.

RAM EFFECT – The effect which causes an increase of temperature and pressure of the air that enters the compressor of an aircraft gas turbine unit due to aircraft speed. Sometimes called RAM.

RAM EFFICIENCY – Actual pressure rise realized in a diffuser compared with the pressure rise possible under reversible conditions. Also called INTAKE EFFICIENCY.

REACTION TURBINE – The turbine in which the nozzles and moving blades are each made in the same general form, with the cross-section in the direction of the gas flow reduced so that both the fixed and moving blades act as expanding nozzles.

REGENERATIVE PLANT – The plant that utilizes a heat exchanger to recover heat from the turbine exhaust gases and thereby decreases the heat required to be added in the combustor.

REHEAT CYCLE – The unit in which the expansion of the hot gases is carried out in two stages, and reheating of the working fluid to the upper limit of temperature takes place between the stages of expansion.

ROCKET – A self propelled unit in which the fuel and the oxidant are contained within the shell.

SECONDARY AIR – Air quantity which is about five times the minimum air for combustion, that flows around the annular space of the combustion chamber and cools the products of combustion.

SECONDARY ZONE – Portion of the combustion chamber wherein about 30% of air is added at the right points in the combustion process so as to complete the combustion of fuel.

SHAFT, ROTOR, SPINDLE – The rotating assembly of the turbine which carries the blades.

SHROUD – A band placed around the periphery of the blade tips in order to stiffen the blades and prevent spillage of the fluid over the blade tips. Shrouds may be continuous or in segments integral with one or more blades.

SINGLE SHAFT GAS TURBINE – Turbine unit in which all compressors and turbines in the plant are mounted on one shaft and are coupled in series arrangement.

SLIP – The failure of the whirl velocity of air at the outlet of the centrifugal impeller becoming equal to the impeller tip speed.

SLIP FACTOR – Ratio of whirl velocity of air at the outlet of the compressor (centrifugal type) to the impeller tip speed. This factor limits the work capacity of the compressor even under isentropic conditions.

SOLIDITY OF A BLADE – The ratio of blade chord to pitch, and ranges from 1.0 to 2.0.

SONIC VELOCITY – Speed of sound in a gas. This is the velocity at which a pressure wave is propagated throughout the gas and this velocity depends upon the pressure and density of the gas.

SPECIFIC POWER – Net horse power developed per kg per second of air flow.

STAGGER ANGLE – The angle made by the axial direction and the chord line, a parameter describing the setting of a row of blades of given form and spacing.

STAGE OF A TURBINE – Unit which consists of the fixed nozzle row (or fixed blade row) and the moving row of blades which receives the gases.

STAGNATION ENTHALPY – Enthalpy of a moving gas when brought to rest isentropically. Also called TOTAL HEAD.

STAGNATION TEMPERATURE – The hypothetical temperature which would result if all the kinetic energy of a flowing gas were to be converted into heat under conditions of no gain or loss of heat *i.e.*, adiabatic conditions. Also called TOTAL HEAD TEMPERATURE. This is the sum of static temperature and dynamic temperature.

STALLING – The phenomenon of reduction in the lift force (acting on an air foil) at higher angles of incidence. Also called LIMIT OF STABILITY.

STATIC HEAD EFFICIENCY – Ratio of the temperature equivalent of the work output and leaving energy to the isentropic temperature drop from the total head inlet to the static outlet pressure.

STEADY FLOW – Flow of fluid in which the quantities such as velocity, pressure, temperature etc., may change from point to point but they do not vary with time at any particular point.

STRAIGHT COMPOUNDED UNIT – The system in which the low pressure compressor is driven by the low pressure turbine and the high pressure compressor by the high pressure turbine. Power is taken from the low pressure turbine shaft.

SUBSONIC DIFFUSER – A diffuser having a diverging cross-section in the direction of flow.

SUPERSONIC DIFFUSER – A diffuser having a converging cross-section in the direction of flow.

SYMMETRIC STAGE AXIAL FLOW COMPRESSOR – An axial flow compressor which has symmetric blade arrangement so that the pressure rise in the moving row and the pressure rise in the fixed row are equal.

TAPERED BLADES – The turbine blades taper (decrease in depth) from base to tip, so as to diminish the centrifugal stress at the various sections, at the base or hub sections of the blade.

TERTIARY ZONE – Portion of the combustion chamber wherein the left out 50% of air is mixed with the burnt gases so as to cool them down to the temperature suitable to turbine materials.

THRUST OF A JET – Sum total of the pressure thrust and that due to change of momentum.

TOTAL HEAD EFFICIENCY – Ratio of the actual work output to the maximum possible work output that could be obtained with the existing leaving energy.

TOTAL HEAD PRESSURE – Pressure of the moving fluid corresponding to the stagnation or total head temperature.

TURBINE MECHANICAL EFFICIENCY – Ratio of shaft horse power to rotor (internal) horse power.

TURBINE STAGE EFFICIENCY – Ratio of the work delivered to the rotor per unit of gas flowing divided by the isentropic drop available in each unit of gas flowing.

TURBO PROP UNIT – The unit in which the gas turbine drives a propeller and the gases after leaving the turbine are exhausted as a jet to augment the thrust of the propeller.

TWISTED BLADES – Long turbine blades are usually twisted from hub to tip so as to compensate for blade velocity variations and in some cases as well as to satisfy radial pressure equilibrium conditions.

VELOCITY COMPOUNDED IMPULSE TURBINE – The turbine in which the kinetic energy created by the expansion of gas in nozzles is absorbed in two or more rows of moving blades so as to reduce the speed of the turbine rotor.

VOLUMETRIC EFFICIENCY – Ratio of the equivalent volume of free atmospheric air finally delivered by a compressor to the volume of free atmospheric air entering the suction pipe of the compressor.

WINDAGE LOSSES – When moving blades come in contact with inactive fluid, some kind of kinetic energy is imparted to the fluid at the expense of the kinetic energy of the blades. There is also frictional effect. This loss is known as windage losses.

WORK RATIO – The ratio of the actual work output (in heat units) to the isentropic heat drop from the total head inlet to the static outlet conditions.

CHAPTER 20

INTERNAL COMBUSTION ENGINE PARTS

AFTER COOLER – A device used on turbocharged engines to cool air which has undergone compression.

ATDC – After TDC, After top dead centre.

AIR CLEANER – A device mounted on the intake manifold for filtering out unwanted solid impurities such as dirt and dust from air that is being drawn into the engine cylinder through the inlet manifold.

AIR COOLED ENGINE – An engine that is cooled by passage of air around the cylinder, not by passage of a liquid through water jackets.

AIR STANDARD CYCLE – A standard cycle of reference by which the performance of the different internal combustion engines may be compared, and their relative efficiencies calculated.

AKROYD ENGINE – The first compression ignition engine, patented by Akroyd Staurt in 1890.

ALUMINIUM CYLINDER BLOCK – An engine cylinder block cast from aluminium or aluminium alloy, and which usually has cast iron sleeves installed for use as cylinder bores.

ANTIFREEZE – A chemical, added to the coolant (usually ethylene glycol) to lower its freezing point and thereby prevent the coolant from freezing in cold weather.

ANTI ICING SYSTEM – A carburettor unit designed to prevent formation of ice on a surface or in a passage.

ARTICULATED CONNECTING ROD – The auxiliary connecting rods of a radial engine, which work on pins carried by the master rod instead of on the main crankpin. Also called LINK RODS.

BACK PRESSURE – A pressure exerted by a fluid contrary to the pressure producing the main flow. For example, pressure in the exhaust manifold, the higher the back pressure, greater is the resistance to flow of exhaust gases through the exhaust system. This lowers volumetric efficiency.

BARREL – Refers to the cylinders in an engine or to the number of throttle bores in a carburettor.

BARREL TYPE CRANKCASE – A petrol engine crankcase so constructed that the crankshaft can be removed from one end, in more normal construction, the crankcase is split.

BASE CIRCLE – As applied to camshaft, lowest spot on the cam. Area of cam that is directly opposite lobe.

BELLOWS – A device, usually metal that can lengthen or shorten much like an accordian. Some cooling system thermostats are of bellows type.

BIMETAL – A thermostatic bimetal element made up of two different metals with different heat expansion rates. Temperature changes produce a bending or distortion movement of the element.

BLOCK (engine) – Basic part of the engine casting containing cylinders.

BLOWBY – Piston rings do not effectively seal compression pressure, and as such allows hot gases to blow between rings and cylinder wall into the crankcase. This causes overheating of piston and poor performance.

BLOWER – Supercharger or engine intake air compressor, a low pressure air pump, usually rotary or centrifugal type.

BLUE PRINTING (engine) – Dismantling the engine and reassembling it to exact specifications.

BOOST – The amount by which the induction pressure of a super charged internal combustion engine exceeds atmospheric pressure, expressed in kg/sqcm.

BORE – A cylinder, hole, or the inside diameter of the cylinder or hole. May refer to cylinder itself or to diameter of the cylinder.

BORE DIAMETER – Diameter of a hole or a cylinder.

BORING – Renewing or enlarging cylinders by cutting and honing them to a specified size. Boring bar is used to make the cut.

BORING BAR – Tool used to cut engine cylinders to specific size. As used in garages, to cut worn cylinders to a new diameter.

BOTTOM DEAD CENTER (BDC) – Lowest position of the piston in the cylinder.

BOXED I ROD – Connecting rod in which I beam section has been stiffened by welding plates on each side of the rod.

BRAKE HORSE POWER (BHP) – Actual usable power delivered by an engine at the crankshaft for driving a vehicle or any other unit. Computed using the engine coupled to a dynamometer.

BRAKE MEAN EFFECTIVE PRESSURE (BMEP) – Mean effective pressure (imaginary) which when assumed to be acting on the piston during the power stroke would result in the given brake horse power output. Equal to mean indicated pressure times mechanical efficiency.

BRAKE THERMAL EFFICIENCY – Ratio of heat equivalent of power output in the form of brake horse power to the corresponding heat input from fuel.

BREAKIN – Period of operation between installation of new or rebuilt parts and when the parts are worn to the correct fit. Driving at reduced and varying speed for a specified distance or duration permits parts to wear out to the correct fit.

BREATHER PIPE – A pipe opening into the interior of an engine *i.e.*, crankcase. Used to assist ventilation.

BTDC – Before top dead center, also called BUDC-before upper dead center.

CAM – A rotating lobe of irregular shape or eccentric or offset portion of the shaft (cam). It changes rotary motion of cam shaft to reciprocating or variable motion of valve lifter resting on it.

CAM FOLLOWER (valve lifter) – A part which is held in contact with the cam and to which the cam motion is imparted and transmitted to the push rod.

CAM GROUND PISTON – A piston that is ground slightly oval in shape. It becomes round as it expands with heat.

CAM NOSE – Also called CAM LOBE. That portion of the cam which holds the valve wide open. It is the high point of the cam.

CAM SHAFT – The shaft in the engine which has a series of cam lobes (simply called cams) for operating the valve mechanisms, driven by gears or sprockets and chain from the crankshaft.

CAM SHAFT GEAR – The gear that is fastened to the cam shaft.

CAST IN SLEEVE – An aluminium cylinder block cast around an iron cylinder sleeve.

CAST IRON CYLINDER – A one piece cylinder assembly made of cast iron with a machined bore.

CAST PISTON – A piston made by pouring molten aluminium alloy into the mould of desired shape.

CENTRIFUGAL GOVERNOR – A governor which uses flyweight force to sense speed, in order to control the amount of fuel supplied to the combustion chambers.

CENTRIFUGAL OIL SLINGER – Cup shaped centrifugal oil filter mounted to the end of the crankshaft. As oil passes through the slinger, centrifugal force removes impurities that are heavier than oil.

CHROME PLATED RING – A piston compression or oil ring that has its cylinder wall face lightly plated with hard chrome.

CIRCLIP – A circular clip or snap ring that fits into a groove, used to locate or retain a shaft or component.

CLEARANCE – The amount of space between two moving parts or between a moving and a stationary part, such as a journal and a bearing, piston and cylinder.

CLEARANCE VOLUME – The volume remaining above the piston when the piston is at TDC.

CLOSED CRANKCASE VENTILATION SYSTEM – A system in which the crankcase vapours (blow by gases) are discharged into the engine intake system and pass through the engine cylinder rather than being discharged into the air.

COATED BORE – Thin coating of chrome or iron applied to the inside of the cylinder by electroplating, or wire explosion spray coating.

COATED RING – A piston ring having its cylinder wall face coated with ferrous oxide, soft phosphate or tin. This thin coating helps new rings to seat by retaining oil and reduces scuffing during breakin.

COMBUSTION CHAMBER – The space at the top of the cylinder and in the cylinder head or piston or both, in which combustion of fuel and air charge takes place. The space enclosed by the piston, when the piston is at TDC.

COMBUSTION CHAMBER VOLUME – The volume of the combustion chamber when the piston is at TDC, measured in cubic centimeters.

COMBUSTION CYCLE – A series of thermodynamic processes through which the working gas passes to produce one power stroke. The full cycle is intake, compression, power and exhaust.

COMPRESSION CHECK – Measurement of compression pressure in all the cylinders at cranking speed.

COMPRESSION PRESSURE – Pressure in the combustion chamber at the end of the compression stroke, but without any of the fuel being burned.

COMPRESSION RATIO – The ratio between the total volume of the cylinder when the piston is at BDC and the volume when the piston is at TDC.

COMPRESSION RELEASE – A device to prevent the intake valve or exhaust valve from closing completely. This permits the engine crankshaft to be turned over without compression and with ease. Also called DECOMPRESSOR.

COMPRESSION RINGS – The upper ring or rings on a piston designed to hold the compression in the cylinder and prevent or reduce combustion gas leakage *i.e.*, blowby.

COMPRESSION STROKE – The piston stroke from BDC to TDC during which both valves are closed and the charge is compressed into a smaller space creating heat by molecular action.

COMPRESSION TESTER – An instrument for testing the amount of pressure, or compression, developed in an engine cylinder during cranking. Also called COMPRESSION GAUGE.

CONNECTING ROD – The rod made of steel or aluminium alloy usually having an I beam cross-section. A piston pin connects the connecting rod and the piston.

CONNECTING ROD BEARING – Bearings used in the connecting rod small end or big end holes.

CONNECTING ROD CAP – The part of the connecting rod big end assembly that attaches the rod to the crankpin.

CONNECTING ROD TIP – Amount of radial (side) play at the top of the connecting rod. Measurement of rod tip is one way of determining the condition of the rod big end bearing.

CONTROLLED PORT SCAVENGING – Scavenging method using ports which are controlled by valves in addition to the power piston.

COOLANT – The liquid mixture of antifreeze and water circulated in the cooling system of an engine or machinery.

COOLING SYSTEM – In an engine, the system that removes heat by the natural or forced circulation of the coolant and thereby prevents engine overheating. It includes the water jackets, water pump, radiator and thermostat, or cooling fins, blower and cowl.

CORE (**radiator**) – A number of coolant passages surrounded by fins through which air flows to carry away heat from the coolant.

COUNTER FLOW CYLINDER HEAD – has the intake and exhaust passages on the same side of the cylinder head.

COUNTER WEIGHT – Weights that are mounted on the crankshaft opposite each crankthrow. These reduce the vibration caused by the crank and also reduce bearing loads due to inertia of the moving parts.

CRANK – A device for converting reciprocating motion into rotary motion or vice versa.

CRANKCASE – The lower part of an engine in which the crankshaft rotates. It consists of the lower section of the cylinder block, and the oil pan.

CRANKCASE VENTILATING SYSTEM – The system that permits air to flow through the engine crankcase when the engine is running to carryout the blowby gases and relieve any pressure build up.

CRANKPIN – That part of the crankthrow of the crankshaft to which the connecting rod is attached.

CRANKPIN RIDGING – A type of crankpin failure typified by deep ridges worn into the crankpin bearing surfaces.

CRANKSHAFT – The main rotating member, or shaft running along the length of the engine. Portions of the shaft are offset to form throws or cranks to which the connecting rods are attached. Crankshaft is supported by mainbearings.

CRANKSHAFT AXLES – Extension at each end of the crankshaft to provide a mounting place for main bearings, and alternator rotor of magneto flywheel.

CRANKSHAFT GAUGE – A special type of micrometer which can measure crankshaft wear without removing the crankshaft from the block.

CRANKSHAFT GEAR – A gear or sprocket, mounted on the front of the crankshaft. Used to drive the camshaft gear or chain.

CRANKSHAFT WHEEL – Portions of an assembled crankshaft, in the form of wheels that provide a mounting place for crankpin and crankaxles.

CRANK THROW – One crankpin with its two webs (the amount of offset of the journal).

CRANK WEB – The portion of the crank throw between the crankpin and main journal. This makes up the offset.

CRITICAL SPEEDS – Speeds at which the frequency of the power strokes synchronize with the crankshafts natural frequency. If the engine is operated at one of its critical speeds for any length of time, a broken crankshaft may result.

CROSS FLOW CYLINDER HEAD – has the intake and exhaust lines on opposite sides of the cylinder head.

CYCLE – A series of events which continuously repeat in definite order. In an engine, the cycle constitutes the four operations that complete the working process and produce power.

CYLINDER – A round hole or tubular shaped structure in a block or casting in which a piston reciprocates.

CYLINDER BLOCK – The basic framework of the engine to which the other engine parts are attached. It includes the engine cylinders and the upper part of the crankcase.

CYLINDER BORE – Diameter of cylinder opening.

CYLINDER BORING – Bore diameter in a cylinder machined to accept oversize piston. This renews a worn cylinder.

CYLINDER DEGLAZING – Use of a hone to slightly roughen the cylinder walls. It produces a cross hatch pattern which aids in seating of new rings.

CYLINDER HEAD – The part that encloses the cylinder bores, used to cover tops of cylinders. Metal section bolted on top of block. It contains the water jackets, and on I head engines, the valves. Also forms part of combustion chamber.

CYLINDER HONE – An expandable rotating tool with abrasive fingers turned by an electric motor, used to clean and smooth the inside surface of a cylinder to exact measurements.

CYLINDER LEAKAGE TESTER – A type of cylinder tester that forces compressed air into the cylinder through the sparkplug hole when the valves are closed and the piston is at TDC on the compression stroke. The percentage of compressed air that leaks out is measured, and the source of leakage accurately pin points the defective part.

CYLINDER SLEEVE – A replaceable sleeve, or liner, put into the cylinder block to form the cylinder bore. It is either pressed or pushed into the cylinder block.

DEAD CENTER – Point at which the piston reaches its uppermost or lowermost position in the cylinder. At these positions, at the end of the stroke, the crank and connecting rod are in a straight line.

DECARBONIZE – To remove carbon build up on piston, combustion chamber and other parts.

DECOMPRESSOR – is the device that opens the engine intake or exhaust valve and retains it in the opened position. The compression effect is thus reduced and helps easy rotation of the crankshaft at the time of starting.

DIPSTICK – The metal stick that passes into the oil sump. Used to determine quantity of oil in the engine sump.

DIRECT COOLED PISTON – A piston which is cooled by the internal circulation of a liquid.

DISPLACEMENT – In an engine, the total volume of fresh charge an engine is theoretically capable of drawing into all cylinders during one operating cycle. The space swept through by the piston in all cylinders in moving from one end of a stroke to the other.

DOHC ENGINE – An engine having two camshafts over each line of cylinders, one operating intake valves, and the other operating exhaust valves. Double overhead camshaft engine.

DRY SLEEVE – A cylinder sleeve (liner) where the sleeve is supported in the cylinder block metal over its entire length. The coolant does not touch the sleeve itself.

DYNAMOMETER – A device for absorbing and measuring the power output, or brake horse power, of an engine. May be an engine dynamometer, which measures power output at the flywheel, or a chassis dynamometer, which measures the power output at the driven wheels.

ELEMENT FILTER – A disposable oil or air filter that uses gauze or paper as filtering material.

ENBLOCK – One piece, such as an engine cylinder block cast in one piece.

ENGINE – A machine that converts heat energy into mechanical energy (mechanical action in a car). The assembly that burns fuel to produce power, sometimes referred to as the power plant.

ENGINE DISPLACEMENT – Volume of space through which head of piston moves in full length of its stroke multiplied by the number of cylinders in the engine. Result is given in cubic centimetres.

ENGINE TUNE UP – The procedure for checking and adjusting the various engine components so that the engine is restored to top operating condition.

ETHYLENE GLYCOL – A chemical compound (solution) added to the engine coolant to reduce its freezing point and thereby protect the cooling system against freezing of the coolant.

EXHAUST CUTOUT – Y shaped device placed in the exhaust pipe of an engine ahead of muffler. Driver may channel exhaust through the muffler or into the other leg of the Y where the exhaust gases pass out without going through the muffler.

EXHAUST MANIFOLD – A housing with a series of connecting pipes between the exhaust ports and the exhaust pipe through which hot burned gases from the engine cylinders flow.

EXHAUST PIPE – Pipe connecting exhaust manifold to the muffler.

EXHAUST PORT – The opening through which exhaust gases pass from the cylinders to the manifold.

EXHAUST STROKE – The piston stroke from BDC to TDC during which the exhaust valve is open so that the burned gases are forced out from the cylinder.

EXHAUST SYSTEM – A group of parts consisting of the exhaust manifold, exhaust pipe, muffler, tailpipe and resonator if used.

EXHAUST TUNING – Cutting exhaust pipe to such length that provides maximum efficiency.

EXHAUST VALVE – The valve which opens to allow burned gases to exhaust from the engine cylinder into the exhaust manifold during the exhaust stroke.

EXPANSION PLUG – A plug that is slightly dished out and used to seal core passages in the cylinder block and cylinder head. When driven into place, it is flattened and expanded to fit tightly.

EXPANSION RATIO – Ratio of the total volume when the piston is at BDC to the clearance volume when the piston is at TDC (normally equal to compression ratio).

EVAPORATIVE COOLING SYSTEM – A cooling system in which the heat finally passes to atmosphere by evaporation of the coolant. This system may be either open or closed.

FAN (cooling) – The device on the front of the engine that rotates to draw cooling air through the radiator or around the engine cylinders.

FAST IDLE – Engine idle speed when the carburettor fast idle cam is in operation. A mechanism on the carburettor, connected to the automatic choke, that holds the throttle valve slightly open when the engine is cold so that the engine will idle at the higher rpm as long as the choke is applied.

F HEAD ENGINE – A type of engine in which some of the valves are in the cylinder head and some in the cylinder block, giving the F shaped appearance.

FILTER – That part in the air, lubricating oil or fuel system through which air, oil or fuel must pass so that dust, dirt or other contaminants are removed.

FINS (engine) – Thin metal projections on an air cooled engine cylinder and head, which greatly increase the heat radiating surfaces and help cooling of the engine cylinder.

FINS (radiator) – Thin metal projections, over which cooling air flows, that carry heat away from the hot coolant passages to the passing air.

FIRING ORDER – The numerical order in which the engine cylinders fire, or deliver their power strokes begining with No.1 cylinder.

FLYWHEEL – A heavy rotating metal wheel attached to the crankshaft which helps even out the power surges from the power strokes and also serves as part of the clutch and engine cranking system. Acts as power reservoir.

FLYWHEEL RING GEAR – A gear fitted around the flywheel that is engaged by the teeth on the starting motor drive to crank the engine.

FORGED PISTON – A piston made by hammering hot aluminium into a mould of desired shape.

FOUR STROKE CYCLE – The four piston strokes of intake, compression, power and exhaust which make up the complete cycle of events in the four stroke cycle.

FRICTION HORSE POWER – The power used up by an engine in overcoming its own internal friction, usually, it increases as engine speed increases.

FUEL TANK – The storage tank for fuel on the vehicle.

FULL FLOATING PISTON PIN – is one which is free to rotate both in the piston pin bosses and in the connecting rod small end.

FULL FLOW FILTER – Type of oil filter in which all the oil from the oil pump flows through the filter.

GASKET – A flat strip, usually of cork or other material, placed between two non-moving, machined surfaces to provide a tight seal between them and thereby prevent leakage.

GASKET CEMENT – A liquid adhesive material, or sealer applied on gaskets, in some applications, the liquid layer of gasket cement is used as the gasket itself.

GLAZE (cylinder) – The mirror like, very smooth finish that develops on engine cylinder walls during engine operation.

HEAT DAM – In a piston top land a groove cut out to reduce the size of the heat path, allowing the piston skirt to run at lower temperature.

HEAT LAND RING – A compression ring having the cross-sectional shape of the letter L, used as top ring.

HELICOIL – A rethreading device to repair worn or damaged threads. It is installed in a retapped hole to bring the screw thread down to the original size.

HORSE POWER (hp) – A measure of the mechanical power, or the rate at which work is done. One horse power equals 4500 mkg of work per minute.

IDLE SPEED – The speed at which an engine runs without load when the accelerator pedal is released.

I HEAD ENGINE – An overhead valve (OHV) engine with the valves in the cylinder head.

IMPELLER – Finned wheel that produces pressure and flow when spun in an enclosed housing of oil pump or water pump.

INDICATED HORSE POWER (IHP) – The power produced within the engine cylinders before deducting any frictional loss.

INERTIA – Tendency of a stationary object to resist movement or tendency of a moving object to continue moving in same direction.

IN LINE ENGINE – An engine in which all engine cylinders are in a single row, or line.

INTAKE MANIFOLD – is a casting attached to the cylinder head in the case of a overhead valve engine or to the cylinder block in the case of a side valve engine. Through the intake manifold fresh charge enters the cylinders.

INTAKE STROKE – The piston stroke from TDC to BDC during which the intake valve is open and the cylinder receives a charge of air fuel mixture in a SI engine or air alone in a CI engine.

INTAKE VALVE – The valve that opens to permit fresh charge to enter the cylinder on the intake stroke.

INTERNAL COMBUSTION ENGINE (IC engine) – An engine in which the fuel is burnt inside the engine.

JACKETS – The water jackets that surround the cylinders through which the coolant passes.

KINETIC ENERGY – Energy associated with motion. An internal combustion engine produces kinetic energy (crankshaft rotation).

L HEAD ENGINE – A type of engine with valves located in the cylinder block, the combustion chamber is L shaped.

LIGHT LOAD TEST – The test applied to storage batteries during which the voltage is measured while the battery is subjected to a light load, such as the car head lights.

LIQUID COOLED ENGINE – An engine that is cooled by the circulation of the liquid coolant around the cylinders.

LOAD TEST – A cranking motor test to measure the current drawn under normal cranking load.

MAIN BEARINGS – The cylinder block and crankcase unit is provided at the bottom with split main bearings for supporting the crankshaft journals.

MANIFOLD – Intake manifold or exhaust manifold. A casting connecting a series of outlets to a common opening.

MECHANICAL EFFICIENCY – In an engine, the ratio between brake horse power and indicated horse power.

MUFFLER – In the exhaust, a device through which the exhaust gases must pass and which muffles the sound.

NO LOAD TEST – A cranking motor test in which the cranking motor is operated without load, the current drawn and the armature speed at the specified voltage are noted.

OIL CONTROL RING – Piston ring designed to remove excess oil from the cylinder wall, usually bottom ring.

OIL SUMP – is fastened to the bottom of the crankcase. This protects the engine from below and is used as a reservoir for lubricating oil in a four stroke engine.

OPPOSED CYLINDERS – Cylinders positioned opposite each other in the same plane.

O RING – A circular cross-sectional sealing ring, which is compressed into the groove to provide sealing action. Seal used in dynamic application where little or no rotational motion occurs. Also used as a static seal.

OTTO CYCLE – The four operations, namely intake, compression, power and exhaust form a cycle. Named after the inventor Nikolaus Otto and is the basic cycle for all SI engines.

OVERCHARGING – Continued charging of a battery after it has reached a charged condition. This action damages the battery and shortens its life.

OVERHEAD CAMSHAFT (OHC) ENGINE – An engine in which the camshaft is located in the cylinder head or heads instead of in the cylinder block.

OVERHEAD VALVE (OHV) ENGINE – An engine in which the valves are mounted in the cylinder head above the combustion chamber, the camshaft is usually mounted in the cylinder block, and the valves are actuated by push rods.

OVER SQUARE ENGINE – An engine which has a bore larger in dimension than the length of the stroke.

PANCAKE ENGINE – An engine with two rows of cylinders which are opposed and on the same plane, usually set horizontally in a vehicle.

PISTON – A cylindrical part, closed at one end, which moves up and down in the cylinder. Open end is attached to the connecting rod. Combustion pressure is exerted on closed end of piston, causing connecting rod to move and crankshaft to turn.

PISTON CROWN – Top of piston, directly exposed to combustion pressure and heat.

PISTON DISPLACEMENT – The cylinder volume displaced by the piston as it moves from the bottom to the top of the cylinder during one complete stroke.

PISTON PIN – Also called wrist pin. The cylindrical or tubular metal piece that attaches the piston to the connecting rod.

PISTON PIN BOSS – A strengthened section of piston wall extending to inside of piston crown. It supports piston pin.

PISTON PIN HOLE – Machined hole through piston wall where piston pin and retaining circlips are mounted.

PISTON RINGS – Rings fitted into grooves in the piston. These are two types: Compression rings for sealing the compression into the combustion chamber and oil rings to scrape excess oil off the cylinder wall and thereby prevent it from working up into and burning in the combustion chamber.

PISTON RING COATINGS – of relatively soft substances such as phosphate, graphite, and iron oxide aid effective wear in and prevents rapid wear of the ring.

PISTON RING COMPRESSOR – A special tool used in engine overhaul work to compress the piston rings inside the piston grooves so that the piston and rings assembly may be installed in the engine cylinder.

PISTON RING END GAP – Distance between ends of a piston ring when installed in the cylinder. The clearance is measured with a feeler gauge, keeping the piston at BDC.

PISTON RING GROOVE – Grooves machined in the piston external surface to accept piston rings.

PISTON SEIZURE – Overheating of piston to the point where it will no longer move freely in the cylinder.

PISTON SKIRT – The lower part of the piston below the piston pin hole.

PISTON SLAP – A hollow, muffled, bell like sound made by an excessively loose piston slapping the cylinder wall at dead centre positions.

POPPET VALVE – A mushroom shaped valve, widely used in internal combustion engines.

PORT (cylinder) – In an engine, the valve port or opening in which the valve operates and through which the charge or burned gases pass.

POWER – Rate at which work is done.

POWER PLANT – The engine or power producing mechanism in the vehicle.

POWER STROKE – The piston stroke from TDC to BDC during which the charge burns and forces the piston down so that the engine produces power.

PRESSURE TESTER – An instrument that clamps in the radiator filler neck, and is used to pressure test the cooling system for leaks.

PREVENTIVE MAINTENANCE – The systematic inspection, detection and correction of failures in an engine or in a vehicle, either before they occur, or before they develop into major defects.

PUSH ROD – In the I head engine, the rod between the valve lifter and rocker arm.

RADIATOR – A heat exchanger which reduces coolant temperature in a liquid cooling system.

RADIUS RIDE – In a reground crankshaft, if the radius of the journal, where it comes up to the crank cheek, is not cut away enough, the journal will ride on the edge of the bearing. The contact is called radius ride.

REBORE – To bore out a cylinder larger than its original size.

RECIPROCATING ENGINE – Also called piston engine. An engine in which the piston moves up and down or back and forth, as a result of combustion in the top of the cylinder.

RING EXPANDER – A special tool used to expand piston rings for installation on the piston.

RING LAND – Solid area of piston which supports rings, located between ring grooves.

RING RIDGE – A ridge left at the top of a cylinder as the cylinder wall below it is worn by piston ring movement.

RING RIDGE REMOVER – A special tool used for removing the ring ridge from the cylinder.

ROCKER ARM – In an I head engine, a device that rocks on a shaft or pivots on a stud as the cam actuates the push rod causing the valve to open.

ROD BOLTS – Special bolts used on the connecting rod to attach the cap. Sometimes lock nuts are provided.

ROD SMALL END – The end of the connecting rod through which a piston pin passes to connect the piston to the connecting rod.

ROD BIG END – The end of the connecting rod that attaches around the crankpin.

RPM – Revolutions per minute.

REVERSE FLUSHING – A method of cleaning a radiator or an engine cooling system by flushing in the direction opposite to normal coolant flow.

SCORED – Scratched or grooved, as a cylinder wall may be scored by abrasive particles moved up and down by the piston rings.

SCRAPER – A device in engine service, to scrape carbon etc., from engine block, piston etc.

SCRAPER RING – On a piston, a type of oil control ring designed to scrape excess oil from cylinder wall and cause it to flow down into the crankcase.

SCREEN – A fine mesh screen in the fuel or lubricating systems that prevents large particles (dirt and particles of dust) from entering the system.

SCUFFING – A type of wear of moving parts characterized by transfer of material from one to other part and results in pits and grooves.

SEALER – A thick, tacky compound usually spread with a brush, which may be used as a gasket or sealant, to seal small openings or surface irregularities.

SEAT – The surface on which another part rests, such as a valve seat. Also, applied to the process of a part wearing into fit, for example piston rings seat after a few kilometers of driving.

SEMIFLOATING PISTON PIN – is clamped in the piston pin bosses and free to rotate in the connecting rod small end or clamped in the connecting rod small end and free to rotate in the piston pin bosses.

SEVERE RINGS – Piston rings which exert relatively high pressure against the cylinder walls, some times by the use of an expander spring behind the ring. Such rings can be used in an engine having excessive cylinder wear.

SHROUD – A hood placed around an engine cooling fan to improve fan (cooling) action.

SIAMESED CRANKSHAFT – Crankshaft configuration where two rods are mounted on the same crankpin (journal). One rod is forked and the other rod is mounted on inside of the fork.

SIDE BY SIDE CRANKSHAFT – Crankshaft configuration where connecting rods are mounted next to each other on the same crankpin (journal).

SINGLE PIECE CONNECTING ROD – has small end, rod or shank and big end as one unit. This is used in most of the small two stroke engines meant for two wheelers.

SPRING (valve) – A coiled wire that varies its length by flexing, and twisting.

SPRING FREE LENGTH – The length of the spring when measured without any load on it.

SPRING MECHANICAL PRELOAD – The length or pressure of a spring, measured while it is in the installed condition.

SPRING RATE – The amount of force necessary to compress a spring a specific distance, kilograms per centimeter, to indicate the stiffness or softness of a spring.

SPRING RETAINER – In the valve train, the piece of metal that holds the springs in place and is itself locked in place by the valve spring retainer locks.

SPRING RETAINER LOCK – The locking device on the valve stem that locks the spring retainer in place.

SQUARE ENGINE – An engine having the bore and stroke of equal measurements.

SQUISH – The action (radial inward air motion) in some combustion chambers in which the last part of the compressed charge is pushed, or squirted, out of a decreasing space between the piston and the cylinder head.

STIRLING ENGINE – A typical internal combustion engine in which the piston is moved by the changing pressure of a working gas that is alternately heated and cooled.

STROKE – In an engine, the distance that the piston moves from BDC to TDC or vice versa.

SUMP – A system for storing lubricating oil, either in the crankcase (wet sump) or in a separate tank (dry sump).

SUPER CHARGER – A device in the intake system of an engine which pressurizes the ingoing charge. This increases the mass of charge (air fuel mixture) burned and thus increases engine output. If the supercharger is driven by the engine exhaust gas turbine, it is called TURBOCHARGER.

TANK UNIT – The unit of the fuel indicating system that is mounted in the fuel tank.

TAIL PIPE – carries the exhaust gases from the muffler and exhausts the same into the atmosphere. The tail pipe end is sometime cut on a bias (at an angle) to reduce exhaust noise.

THERMAL EFFICIENCY – Relationship between the power output and the energy in the fuel burned to produce the output.

THERMOSTAT – A temperature sensitive device used in a cooling system to adjust flow of coolant as coolant temperature changes.

THERMOSTATICALLY CONTROLLED AIR CLEANER – An air cleaner which uses a thermostat to control the preheating of intake air.

THROW A ROD – Expression used to designate an engine with a loose, knocking connecting rod bearing, or an engine that has broken a connecting rod and showed it through the cylinder block or oil pan.

TIMING – In the engine, refers to timing of valves, and timing of ignition, and their relation to piston position in the cylinder.

TIMING CHAIN – A chain driven by a sprocket on the crankshaft, that drives the sprocket on the camshaft.

TIMING GEARS – are a pair of meshing gears (one bigger gear mounted on the camshaft and another smaller gear mounted on the crankshaft) meant for driving the camshaft at the required speed ratio by the crankshaft.

TOP DEAD CENTRE (TDC) – The piston position at which the piston has moved to the top of the cylinder and the centre line of the connecting rod is parallel to the cylinder wall.

TORQUE – The twisting force at the end of the crank shaft multiplied by the distance of this force application from the shaft centre, measured in kilogram meters or Newton meters.

TORSIONAL VIBRATION – Back and forth motion around a turning centre. Vibration in a rotary direction that causes a twist-untwist action on a rotating shaft, a rotating shaft that repeatedly moves ahead or lags behind the remainder of the shaft; for example, the actions of a crankshaft responds to the cylinder firing impulses.

TUNED INTAKE SYSTEM – An intake system in which the manifold has the proper length and volume to introduce a ramjet or supercharging effect.

TUNE UP – The procedure of inspection, testing and adjusting an engine and replacing any worn parts to restore the engine to its best performance.

TURBOCHARGER – A supercharger driven by the gas turbine which is operated by the engine exhaust gases.

TWO STROKE CYCLE – The series of events namely intake, compression, power and exhaust all of which take place in two piston strokes. Also called TWO CYCLE in a short form.

V ENGINE – An engine with two banks of cylinders set at an angle to each other to form a V.

VALVE – A device that can be opened or closed to allow or stop the flow of a fluid (liquid or gas or vapour) from one to another place.

VALVE CLEARANCE – The clearance between the rocker arm and the valve stem tip in an overhead valve engine: The clearance in the valve train when the valve, is in the closed position. Also called VALVE LASH.

VALVE FLOAT – The condition that exists when an engine valve does not follow the cam profile, failure of the valve to close at the proper time.

VALVE GRINDING – Refacing a valve in a valve facing machine.

VALVE GUIDE – The cylindrical part in the cylinder block or head in which the valve is assembled and in which valve stem moves up and down.

VALVE LIFTER – Also called lifter, tappet, valve tappet and cam follower. A cylindrical part of the engine, which rests on a cam of the camshaft and is lifted, by cam action, so that the valve is opened.

VALVE LIFTER FOOT – The bottom end of the valve lifter, the part that rides on the cam lobe.

VALVE OVERLAP – Number of degrees of crankshaft rotation through which both the intake and exhaust valves are open together.

VALVE ROTATOR – Device used in place of the valve spring retainer, it has a built in mechanism to rotate the valve slightly each time it opens.

VALVE SEAT – The surface in the cylinder head or cylinder block against which the valve face comes to rest.

VALVE SEAT INSERT – Metal rings inserted in the valve seats, usually exhaust, they are of special metal which can withstand high temperature and exhibit minimum wear at these temperatures.

VALVE SEAT RECESSION – Also known as lash loss, the tendency for the valves, in some engines run on unleaded gasoline, to contact the seat in such a way that the seat wears away, or recesses into the cylinder head.

VALVE STEM – The long, thin cylindrical section of the valve that fits and moves in the valve guide.

VALVE TIMING – The timing of valve opening and closing in relation to piston position in the cylinder.

VALVE TRAIN – The valve operating mechanism of an engine, from the camshaft of the valve.

VEE CYLINDERS – Cylinders positioned at angles to each other forming the shape of the letter V.

VIBRATION DAMPER – is attached to the crankshaft in order to control the torsional vibration caused by the power impulses.

VOLUMETRIC EFFICIENCY – Ratio between the amount of fresh charge that actually enters an engine cylinder and the theoretical amount that could enter under ideal conditions.

WANKEL ENGINE – A rotary type engine, in which a three lobe rotor turns eccentrically in a specially shaped housing.

WET LINER – When fitted in the cylinder block has water on the external surface of the liner. Good cooling is realized by having the water in direct contact with the liner.

WRIST PIN – A cylindrical (solid or hollow) pin that attaches the piston to the connecting rod.

CHAPTER 21

SPARK IGNITION ENGINES

ACCELERATOR – Device for rapid control of speed, for quick opening and closing of the throttle. It is a foot or hand operated, spring returned, linked to the throttle valve in the carburettor. The minimum throttle opening is controlled by the setting of the throttle screw.

ACCELERATOR PUMP – In the carburettor, a small pump linked to the accelerator which momentarily injects a charge of fuel into the intake tract in addition to that supplied by the normal metering components, and thus enriches the mixture when the accelerator pedal is depressed.

ADVANCE – Setting the ignition timing so that spark occurs before the piston reaches top dead center.

AIR BLEED – An opening into a gasoline passage through which air can pass or bleed into the gasoline as it moves through the passage, to weaken the air fuel mixture.

AIR FUEL MIXTURE – Finely atomized mist of fuel and air necessary for combustion. This mixture consists of approximately 15 parts air to one part fuel (15 : 1) at cruising speed.

AIR FUEL RATIO – The proportion of air to fuel in the working charge of an internal combustion engine, or in other combustible mixtures, expressed by weight for liquid fuels and by volume for gaseous fuels.

AIRGAP (spark plug) – Distance between centre and side electrodes, in a spark plug. Spark jumps across this gap.

AIR HORN – In the carburettor, the tubular passage through which the incoming air must pass.

AIR JET – A small jet in the air passage of a carburettor. This jet meters the amount of air fed to the diffuser in an air bleed type carburettor.

ANTIKNOCK SUBSTANCES – Substances added to petrol to lessen its tendency to detonate, or knock in an engine, *i.e.*, Tetra ethyl lead.

ANTIKNOCK VALUE – The relative immunity of a volatile liquid fuel from detonation, or knocking, in a petrol engine, as compared with some standard fuel.

ANTIPERCOLATOR – Device for venting vapours from main discharge tube or well of a carburettor.

ANTI SIPHON SYSTEM – Use of a small passage designed into a carburettor to prevent fuel from siphoning from the float bowl into the engine.

ATOMIZED – Tiny particles of fuel mixed with air, making a fine mist.

AUTOMATIC CHOKE – A carburettor choke device (valve) that automatically positions itself in accordance with carburettor needs or engine temperature.

AUTOVAC – A vacuum operated mechanism for raising fuel from a tank situated below the level of the carbutettor to a position from which it may be fed to the latter by gravity.

BACKFIRE – (1) Premature ignition during starting of an internal combustion engine, resulting in an explosion before the end of compression stroke, and consequent reversal in the direction of rotation. (2) An explosion of live gases accumulated in the exhaust system due to incomplete combustion in the cylinder.

BACKFIRE (intake system) – Preexplosion of air fuel mixture so that the explosion passes the open intake valve and flashes back through the intake manifold. May be caused by faulty timing, crossed plug wires, leaky intake valve etc.

BACK KICK – Violent reversal of an internal combustion engine crankshaft rotation, during starting due to backfire.

BALANCED CARBURETTOR – Carburettor in which the float bowl is vented into the air horn, below the air cleaner, to compensate for the effects of a clogged air filter.

BATTERY COIL IGNITION – High tension supply for sparking plugs, in automobiles, in which the interruption of a primary current from a battery induces a high secondary emf in another winding on the same magnetic circuit, the high potential being distributed in synchronism with the contact breaker in the primary circuit and the engine firing order.

BERNOULLIS PRINCIPLE – Given a fluid flowing through a tube, any constriction or narrowing of the tube will create an increase in the fluid velocity and a decrease in pressure. This principle is used in the venturi tube of the carburettor.

BOOST VENTURI – also called secondary venturi is a smaller venturi or restriction, incorporated in some carburettors in the middle of the primary venturi. It increases air speed, vacuum created and hence fuel flow.

BOWL VENT – is an opening in the carburettor float chamber. This hole prevents pressure or vacuum from building up in the bowl.

BREAKER ARM – The movable arm upon which one of the breaker points of the ignition system is affixed.

BREAKER POINTS (ignition) – Pair of points, one fixed and another movable, that are opened and closed to break and make the primary circuit. When the circuit is broken by opening the points, the spark plug fires.

BUTTERFLY VALVE – A type of valve used for choke and throttle valve in a carburettor that is so named due to its resemblance to the insect of same name. This valve controls charge flow.

CAM ANGLE (ignition) – Number of degrees breaker cam rotates from the time breaker points close until they open again. Also called DWELL ANGLE.

CAPACITOR DISCHARGE IGNITION (CDI) – An electronic ignition system designed to produce very high voltage, consisting of an exciter coil, a capacitor, diode, trigger coil, silicon controlled rectifier and ac ignition coil.

CARBURETION – The actions that take place in the carburettor, converting liquid fuel to vapour and mixing it with air to form a combustible mixture.

CARBURETTOR – The mixing device in the fuel system which meters and mixes gasoline into the air stream (vaporizing gasoline as it does so) in varying proportions to suit engine operating conditions.

CARBURETTOR ADAPTER – Adapter used to fit or place one type of carburettor on an intake manifold that may not be originally designed for it.

CARBURETTOR CIRCUITS – Series of passages and units designed to perform a specific functions—idle circuit, full power circuit etc.

CARBURETTOR ICING – Formation of ice on throttle plate or valve. As fuel nozzles feed fuel into air horn it turns to a vapour. This robs heat from air. When weather conditions are just right (fairly cold and quite humid) ice may form.

CARBURETTOR INSULATOR – A spacer, or insulator, used to prevent excess engine heat from reaching the carburettor.

CENTRIFUGAL ADVANCE (**distributor**) – Unit designed to advance and retard ignition timing through action of centrifugal force resulting from changes in engine speed.

CHOKE – Near the top of the carburettor, a butterfly valve that is closed when starting a cold engine. It chokes off the air flow through the air horn, producing a partial vacuum in the air horn for greater fuel delivery and a richer mixture supply to the engine.

CHOKE STOVE – Heating compartment in or on the exhaust manifold from which hot air is drawn to the automatic choke device.

COIL (**ignition**) – Unit used to step up the relatively low voltage supplied by the battery to the extent necessary to create a spark across the spark plug terminals.

COIL BUILDUP – Build up of a magnetic field while current is flowing through primary windings of the coil.

COLD PLUG – has a shorter heat path. Hence it runs at a much lower temperature than a hot plug.

COMBUSTION LAG TIME – A period of slow burning that occurs before the burning of the air fuel mixture, which spreads throughout the engine combustion chamber.

CONDENSER (**ignition**) – Unit installed between breaker points and coil to prevent arcing at breaker points. Condenser absorbs and retains momentary surge of current when the breaker points open.

CONSTANT CHOKE CARBURETTOR – is the carburettor in which the air and fuel flow passages (*i.e.*, areas) are always maintained to be constant. But the pressure difference or depression which causes the flow of fuel is being varied as per the demand on the engine.

CONSTANT VACUUM CARBURETTOR – is the carburettor in which the air and fuel flow areas are being varied as per the demand on the engine, while the depression or vacuum is maintained to be always same.

CONTACT POINTS – In the conventional ignition system, the stationary and the movable points in the primary circuit, usually made of tungsten, platinum or silver. Also called BREAKER POINTS.

CONVENTIONAL IGNITION – Ignition system which uses breaker points.

CRITICAL COMPRESSION RATIO – The lowest compression ratio at which any particular fuel air mixture will ignite by compression under prescribed test procedure. The lower the critical compression ratio, the better ignition qualities the fuel has (Gasoline engine 4 :1, oil engine 7 :1 diesel engine 12.5 :1).

DASHPOT **(carburettor)** – A device in the carburettor that prevents excessively sudden closing of the throttle.

DETONATION – A violent, instantaneous explosion of the final portion of the burning combustion gases caused by an excessive rise of pressure and temperature, also called AUTO IGNITION.

DIESELING – A condition in which a spark ignition engine continues to run after the ignition is shut off. Also called RUNNING ON.

DISTRIBUTOR **(ignition)** – Unit designed to make and break the ignition primary circuit and to distribute resultant high voltage to the proper spark plug in the cylinder at the correct time. Rarely used in motor cycles.

DISTRIBUTOR CAP **(ignition)** – Insulated cap containing a central terminal with series (one per cylinder) of terminals that are evenly spaced in circular pattern around the central terminal. Secondary high voltage travels to central terminal where it is then channeled to one of the outer terminals by the rotor.

DOWN DRAFT CARBURETTOR – Carburettor air horn is so arranged that the air passes downward through the carburettor on its way into the intake manifold.

DUAL BREAKER POINTS **(ignition)** – Distributor using two sets of breaker points to increase cam angle so that even at high speeds, spark with sufficient intensity will be produced.

DUAL CARBURETTOR – An engine on which two carburettors have been mounted.

DWELL ANGLE – The number of degrees on the breaker cam during which the breaker points are kept closed.

DWELL METER – A device used to measure the number of degrees that the ignition contact points remain closed.

DYNAMIC TIMING – A test of ignition timing made with the strobe light.

ECONOMIZER VALVE – Fuel flow control device within the carburettor.

ELECTRIC ASSIST CHOKE – A choke which uses a small electric heating element to warm the choke spring, causing it to release more quickly. This reduces exhaust emissions during the start up of a cold engine.

ELECTRODE **(spark plug)** – Centre electrode rod passing through the insulator forms one electrode. The rod welded to the shell forms another. They are referred as centre and side electrodes.

ELECTRONIC FUEL INJECTION – A fuel injection system used for injecting gasoline into the spark ignition engines, which has an electronic control system to time and meter the fuel injected.

ELECTRONIC IGNITION SYSTEM – An ignition system using transistors, which does not have mechanical contact breaker points in the distributor, but uses the distributor for distributing the secondary voltage to the spark plugs.

ERODED PISTON **(crown)** – A condition caused by detonation or preignition where the gas temperatures are raised so high that part of piston crown is heated and melted away.

ESC – Electronic spark control.

ENERGY TRANSFER MAGNETO – A self powered ignition system that consists of a flywheel with permanent magnets that rotate around a laminated core with coil windings, an induction coil, contact points and capacitor.

FLAME FRONT – The glowing layer of flame that separates the burned charge from the unburned charge in a SI engine during combustion process. The flame front should move in a controlled pattern across the cylinder.

FLAME VELOCITY – is the speed with which the flame front travels inside the combustion chamber. This affects combustion phenomena, development of pressure and production of power.

FLASH OVER – A condition that occurs when a spark jumps across the surface of a spark plug insulator from the terminal.

FLAT SPOT – A point during acceleration when the engine seems to loose power for an instant.

FLOAT BOWL – In the carburettor, the reservoir from which gasoline feeds into the passing air. Also called FLOAT CHAMBER.

FLOAT LEVEL – The float position at which the needle valve closes fuel inlet to the carburettor to prevent further delivery of fuel.

FLOAT SYSTEM – The system in the carburettor that controls the entry of fuel and fuel level in the float bowl.

FLOODED – A term used to indicate that the engine cylinders received an air fuel mixture too rich to burn.

FOUR BARREL CARBURETTOR – A carburettor with four throttle valves. In effect two, two barrel carburettors in a single assembly.

FUEL FILTER – A screen used to prevent contaminants in the fuel from entering the carburettor or fuel pump.

FUEL LINE – The pipe or tube through which fuel travels from the tank to the fuel pump and from the pump to the carburettor.

FUEL NOZZLE – The tube in the carburettor through which gasoline feeds from the float bowl into the passing air. In a fuel injection system, the tube that delivers the fuel into the compressed air or passing air stream.

FUEL PUMP – The electrical or mechanical device in the fuel system which transfers fuel from the fuel tank to the carburettor.

FUEL SCREW – A fine point screw that projects into the slow jet outlet passage, used to adjust the fuel mixture at slow speeds, located at the engine side of the carburettor.

FUEL SYSTEM – In an automobile, the system that delivers to the engine cylinders, the combustible mixture of vaporized fuel and air. It consists of fuel tank, lines, gauge, carburettor, fuel pump and intake manifold.

FULL ADVANCE – The point at which the advance unit will no longer continue advancing ignition timing.

FULL THROTTLE – A wide open throttle position with the accelerator pressed all the way down to the floor board.

HEAT RANGE – A term used to describe the ability of a spark plug to carry away heat. Plugs with longer nosed insulators take longer path and time to carry heat off effectively.

HEMISPHERICAL COMBUSTION CHAMBER – A combustion chamber shaped like a round dome, allowing use of large valves placed opposite each other in the chamber.

HIGH SPEED CIRCUIT – The circuit in the carburettor that supplies fuel into the air passing through the air horn during, medium and high speed, part to full throttle operation.

HIGHEST USEFUL COMPRESSION RATIO – is the compression ratio at which a fuel test engine can be operated without detonation with any mixture strength or with any ignition timing, at a speed of 1500 rpm.

HOLED PISTON – A condition caused by severe detonation or preignition, where a hole is eaten or burned through the piston crown, as a result of extreme heat and pressure.

HOT PLUG – has a longer heat path, hence it runs at a much higher temperature than a cold plug.

IDLE MIXTURE – The air fuel mixture supplied to an engine during idle. The mixture is usually rich. The idle mixture screw(s) are sometimes adjusted as a part of tune up.

IDLE MIXTURE ADJUSTMENT SCREW – The adjustment screw in some carburettors, that can be turned in or out to vary the quality of the idle mixture.

IDLE SYSTEM – In the carburettor, the passages through which fuel is fed when the engine is idling.

IGNITION (spark) – In an engine, the act of spark in starting the combustion process in the cylinder.

IGNITION ADVANCE – To set the ignition timing, so that a spark occurs earlier or more degrees before TDC.

IGNITION COIL – That part of the ignition system which acts as a transformer to step up the battery voltage to many thousand volts, the high voltage surge then produces a spark at the spark plug gap.

IGNITION DISTRIBUTOR – That part of the ignition system, which opens and closes the circuit to the ignition coil with correct timing and distributes to the proper spark plugs the resulting high voltage surges from the ignition coil.

IGNITION SWITCH – The switch in the ignition system which is operated with a key to open and close the ignition primary circuit.

IGNITION SYSTEM – The part of the electrical system that provides high voltage sparks to the engine cylinder to fire the compressed air fuel mixture, consisting of a voltage source, timing device, capacitor, ignition coil, secondary wiring and spark plugs.

IGNITION TIMING – is the crank angle at which spark occurs relative to the top dead center, during compression stroke.

INTAKE MANIFOLD (SI engine) – The part of the engine that provides a series of passages from the carburettor to the engine cylinders through which air fuel mixture can flow.

INTAKE STROKE – The piston movement from TDC to BDC that occurs as the intake valve opens. This movement causes entry of fresh charge.

JET – A calibrated passage in the carburettor through which fuel flows.

KNOCKING COMBUSTION – is the autoignition or instantaneous ignition of the end charge due to the compression of the same by the expansion and radiation heat of the burning charge.

LEAN MIXTURE – An air fuel mixture that has a relatively high proportion of air and a low proportion of fuel.

LOW SPEED CIRCUIT – The circuit in the carburettor that supplies fuel to the air passing through the air horn during low speed, part throttle operation.

MAIN FUEL NOZZLE – The fuel nozzle in the carburettor venturi that supplies fuel when the throttle is opened partially to fully opened position.

MANIFOLD VACUUM – The vacuum in the intake manifold that develops as a result of the vacuum in the cylinders on their intake strokes.

METERING ROD AND JET – A device, consisting of a small movable rod, which has a varied diameter, and a jet (which accommodates the movable rod) that increases or decreases fuel passage and hence the flow of fuel according to engine throttle opening, engine load or a combination of both.

MISSING – In the engine, the failure of the air fuel mixture in a cylinder to ignite when it should and thus causing the engine to run roughly.

MULTIPOINT FUEL INJECTION (MPFI) – has one injector for each cylinder. Fuel is injected in more than one location. This is often called port injection.

NORMAL COMBUSTION – refers to the combustion of the entire air fuel mixture in the SI engine combustion chamber, layer by layer by the moving flame.

OCTANE NUMBER OF A FUEL – is the percentage by volume of iso-octane in a mixture of iso-octane and *n*-heptane which will exhibit the same antiknock characteristic of the fuel under test, when tested in a standard (CFR) engine, under a set of standard test conditions.

OCTANE RATING – The number indicating the quality of gasoline based on its ability to resist knock. The higher the number, the better the quality. Higher compression engines require higher octane fuel.

OPTIMUM SPARK TIMING – is the ignition timing which will cause that half of the pressure rise occurs at dead center. This happens in practice when 75 per cent of the charge burns after ignition.

PERCOLATION – A condition in which the fuel actually "boils" due to excess heat. Percolation prevents proper atomization of the fuel causing rough running.

PERFORMANCE NUMBER – is the ratio of knock limited imep with the fuel in question to the knock limited imep with iso-octane when the inlet pressure is used as the dependent variable.

PING – A metallic ratling sound produced by the engine under acceleration. It is usually due to incorrect ignition timing or poor grade of gasoline.

POST COMBUSTION ACTIVITY – The last phase of combustion, during which the piston descends, the volume inside the cylinder increases, and the cylinder eliminates spent gases.

POWER PISTON – In some carburettors, a vacuum operated piston that allows additional fuel to flow at wide open throttle to permit delivery of a richer air fuel mixture to the engine.

PREIGNITION – Abnormal combustion that occurs when the air fuel mixture ignites before the spark plug fires and this is due to some hot spot.

PRIMARY IGNITION CIRCUIT – Section of the ignition circuit including contact points, condenser and primary winding of ignition coil, power supply (battery or ignition generating coil), ignition switch and related wiring.

PRIMARY WINDING – Part of an ignition coil, a separate winding of heavy wire wound a few hundred turns around a laminated steel core.

QUENCH – The space in some combustion chambers which absorbs enough heat to quench or extinguish the combustion flame front as it approaches a relatively cold cylinder wall. This prevents detonation of the end gas but results in hydrocarbon emission.

REED VALVE – A type of valve used in the crankcase of some two cycle engines. Air fuel mixture enters the crankcase through the reed valve, which then closes as pressure builds up in the crankcase.

RESISTOR SPARK PLUG – A spark plug incorporating a resistor to shorten the spark duration. This suppresses radio interference and lengthens plug life.

RICH MIXTURE – An air fuel mixture with high proportion of fuel.

RISE TIME – The length of time between the begining of the voltage at the plug and the begining of the spark.

RUN ON – Condition when a SI engine continues to run, even when the ignition key is turned off. Also called DIESELING.

SECONDARY IGNITION CIRCUIT – Part of the ignition circuit consisting of secondary windings of ignition coil, spark plug wire, spark plug terminal and spark plug.

SECONDARY WINDING – Part of an ignition coil, a winding of fine wire wound many thousands of turns around a laminated steel core.

SIDE DRAUGHT CARBURETTOR – also called horizontal carburettor, consists of a horizontal mixing tube, with the float chamber on the side of it.

SINGLE POINT INJECTION – has one or two injectors mounted inside the throttle body assembly. Fuel is sprayed into one point or location at the centre of the engine intake manifold. Single point injection (SPI) is also called throttle body injection (TBI).

SPARK ADVANCE – The adjustment in ignition timing made by the timing device for changes in load, speed or other conditions.

SPARK DURATION – The length of time spark occurs at a spark plug.

SPARK PLUG – The assembly, which includes a pair of electrodes and an insulator, that has the purpose of providing a spark gap in the engine cylinder.

SPARK PLUG HEAT RANGE – The distance heat must travel from the centre electrode to reach the outer shell of the plug and enter the cylinder head.

SPARK PLUG REACH – is the length of the threaded portion of the spark plug. This is the distance between the end of the plug threads and the seat or sealing surface of the plug.

SQUISH – is the radial inward or transverse gas motion that occurs towards the end of the compression stroke, when a portion of the piston head approaches the cylinder head closely.

STATIC TIMING – The test of ignition timing made with a buzzbox or a continuity light while the engine is at rest.

STRATIFIED CHARGE ENGINE – In a gasoline fuel, spark ignition engine, a type of combustion chamber in which the flame starts in a very rich pocket or layer of fuel air mixture and after ignition, spreads to the leaner mixture filling the rest of the combustion chamber. The diesel engine is a stratified charge engine.

SURFACE IGNITION – is the ignition of the air fuel mixture by any hot spot in the combustion chamber.

S/V RATIO – The ratio of surface area of the combustion chamber to its volume, with the piston at the top dead centre. Often used as a comparative indicator of hydrocarbon (HC) emission levels from an engine.

THROTTLE (valve) – A round disc valve in the throttle body of the carburettor that can be turned by the driver to admit more or less air fuel mixture, thereby control the engine speed.

THROTTLE CRACKER – Linkage from the starting motor switch to the throttle, which opens the throttle slightly when the engine is being cranked.

THROTTLE RETURN CHECK – A device in the carburettor which prevents excessively sudden closing of the throttle, also called dashpot.

TIMING LIGHT – A light that is connected to the ignition system to flash each time the number one spark plug fires, used for adjusting the timing of the ignition spark.

TIMING UNIT – A device that determines, when the ignition system fires the spark plugs.

TRANSISTOR ASSISTED CONTACT IGNITION (TAC) – An ignition system similar to battery point ignition but which uses a transistor to eliminate heavy current flow at the contact points.

TRANSISTOR POINTLESS IGNITION – A battery assisted ignition system using a pulse generator to trigger a transistor to fire the spark plug.

TURBULENCE – The state of being violently disturbed. In the engine, the rapid swirling motion imparted to the air fuel mixture entering the cylinder.

TURNS RATIO – The ratio of the number of coils in the primary and secondary windings of an ignition coil.

TWO BARREL CARBURETTOR – A dual carburettor in which there are two throttle valves.

UNIT DISTRIBUTOR – An ignition distributor, used by General Motors, that uses a magnetic pick up coil, and timer core instead of points and condenser. It has the ignition coil assembled in the distributor as a unit.

UPDRAFT CARBURETTOR – is the carburettor in which the air fuel mixture flows upward. This unit can be placed on the side of the engine.

VACUUM ADVANCE – A method of advancing the ignition timing by applying engine vacuum to a diaphragm mounted on the distributor.

VACUUM GAUGE – In automotive engine service, a device that measures intake manifold vacuum and thereby indicates action of engine components.

VAPOUR LOCK – A condition in the fuel system in which gasoline has vapourized and turned to bubbles in the fuel line, or fuel pump so that fuel delivery to the carburettor is prevented or retarded.

VAPOURIZATION – is the process of changing the liquid or atomized fuel into a vapour.

VENTURI – In the carburettor, the restriction in the air horn that produces the vacuum responsible for the movement of gasoline into the passing air.

VENTURI PRINCIPLE – The physical law which states that air moved through a constriction increases in speed and decreases in pressure at the point of constriction.

WEDGE COMBUSTION CHAMBER – A combustion chamber resembling in shape, a wedge.

CHAPTER 22

COMPRESSION IGNITION ENGINES

ACCUMULATOR – A device used for storing liquid under pressure (sometime used to smooth fluid flow).

ADVANCE **(injection timing)** – To set the timing of the injection pump or injectors for an earlier injection.

AFTER BURNING – The burning of fuel that is left in the combustion space when the fuel injection stops.

AFTER COOLER – A device used on turbocharged engines to cool the air which has undergone compression, before its entry into the engine cylinder.

AIR CELL – A small auxiliary combustion chamber used in certain types of compression ignition engines, for promoting turbulence and improving combustion.

AIR FUEL RATIO – At full load operation, the air fuel ratio is at least 35 to 50 per cent greater than the stoichiometric value. Air fuel ratio in a normal diesel engine varies from around 100 : 1 at idle speed to about 30 : 1 at full load.

AIR INJECTION SYSTEM – The system which injects the required quantity of fuel into the combustion space with the aid of compressed air.

AIRLESS INJECTION – Injection of liquid fuel into the cylinder of an oil engine by a high pressure fuel pump, so dispensing with the compressed air necessary in the early diesel engines. Also called SOLID INJECTION or MECHANICAL INJECTION.

AIR STARTING VALVE – A valve which admits compressed air to the air starter for starting purposes.

AIR VENT – Arrangement that helps to remove air from the fuel injection system.

ANTECHAMBER – A small auxiliary combustion chamber, used in some compression ignition engines, in which partial combustion of fuel takes place and this is used to force the burning mixture into the cylinder, so promoting more perfect combustion.

ATOMIZATION – The breaking up of fuel jet into fine particles as it is sprayed into the combustion chamber.

ATOMIZER – A device which disperses liquid fuel into fine particles (pulverized spray).

BIFUEL ENGINE – has two injectors to inject two fuels. In this a small amount of a suitable auxiliary fuel is injected into the cylinder either during the intake stroke or early in the compression stroke. Slightly latter in the stroke, the primary fuel is injected.

BLUE SMOKE – The smoke that results from the burning of lubricating oil that reaches the combustion chamber.

BOSCH METERING SYSTEM – A fuel metering system in a diesel engine, with a helical groove in the plunger which covers and uncovers ports in the pump barrel and thereby varies the effective stroke of the fuel pump.

CAVITATION – The formation of cavities in the fluid due to excessive speed of the activator resulting in loss of efficiency in the pump.

CHEMICAL DELAY PERIOD – The time that elapses between the begining of chemical reaction and the begining of ignition.

CLOSED TYPE NOZZLE – A hydraulically operated, spring loaded needle valve, which opens inward under the pressure acting on the differential area of the needle valve (which is a cylinder lapped in with the body and seated by a spring when the fuel pressure is reduced sufficiently).

COLD SMOKE – The smoke that is made up of droplets of unburned or partly burned fuel or due to water vapour. Also called WHITE SMOKE.

COMBUSTION CHAMBER – The space wherein combustion of fuel with air takes place, more or less equal to the clearance volume.

COMBUSTION SWIRL – Air motion created by the ejection of the combustion products from the precombustion chamber into the clearance space above the piston.

COMMON RAIL SYSTEM – The fuel injection system which consists of a high pressure pump which distributes the fuel to a common rail or header to which injectors are connected.

COMPRESSIBILITY – The property of a substance by virtue of which its density increases with increase in pressure.

COMPRESSION IGNITION – Ignition of fuel due to the heat of compression.

COMPRESSION SWIRL – Rotary motion given to air, as the air is forced, during compression stroke, from the cylinder into the spherical or cylindrical combustion chamber through the throat which is located tangential to the combustion chamber.

CONSTANT PRESSURE COMBUSTION – Combustion which occurs without a change in pressure. In an engine, this is obtained by a slower rate of burning than with constant volume combustion.

CONTROLLED COMBUSTION – During uncontrolled combustion, high temperature and pressure prevail within the combustion chamber. After this combustion, fuel that is injected burns without any delay. By controlling the rate of injection, complete control is possible over the rate of burning.

CONTROL RACK – A toothed rack that runs along the upper end of the fuel pump and engages pinions (gear teeth) on each pump plunger to control the amount of fuel injected and thereby determines engine power output.

CRITICAL COMPRESSION RATIO – Lowest compression ratio at which any particular fuel will ignite by compression under prescribed test procedure. The lower the critical compression ratio the better ignition qualities the fuel has.

DEGREE OF ATOMIZATION – is indicated by the smallness of the size of the particles in the spray and also by the smallness of the variation in the size of the particles.

DELAY PERIOD – Time interval between the start of injection and begining of combustion as indicated by a rise in the pressure crank angle curve, from the curve which represents compression and expansion of air while motoring. Also called IGNITION DELAY.

DELIVERY VALVE – A spring loaded valve mounted at the top of the fuel injection pump barrel. This helps to cut off fuel spray from the nozzle tip abruptly and without dribble.

DEPTH FILTER – Fuel filter which has dozens of layers of porous material arranged in sequence and this arrangement compels the contaminant particles to follow tortuous path into the media.

DIESEL CYCLE – An engine cycle of events in which air alone is compressed and fuel oil is injected at the end of the compression stroke. The heat produced by compressing the air ignites the fuel oil, eliminating the need for spark plugs or a spark ignition system.

DIESEL ENGINE – An engine that operates on the diesel cycle and burns diesel oil by the heat of compression of air.

DIESEL INDEX – A rating of fuel according to its ignition qualities. The higher the diesel index number, the better is the ignition quality of fuel.

DIESEL KNOCK – Sudden, steep pressure rise due to instantaneous uncontrolled combustion of the fuel that has got accumulated during delay period. The pressure wave hitting piston and cylinder walls produce knocking sound.

DIRECT INJECTION ENGINES – have a single open combustion chamber into which the entire quantity of fuel is injected directly.

DISPERSION OF SPRAY – The divergency of the fuel spray which helps distribution of fuel droplets in air as uniform as possible.

DRIBBLING – Unatomized fuel running from the fuel nozzle.

DUAL FUEL ENGINE – An engine that uses an air gas mixture which is being ignited by a small quantity of fuel oil injected at the end of the compression process.

DURATION OF FUEL INJECTION – The crank angle during which fuel is injected into the engine cylinder. Usually as load increases this duration has to be increased.

EFFECTIVE STROKE OF PLUNGER – The distance from the point of closure of the ports by the plunger top to the point of opening of the spill port by the helical groove on the surface of the plunger. This decides duration and quantity of fuel delivery by the pump.

ENGINE DERATING – Reducing maximum fuel flow to the engine.

FUEL FILTER – A kind of strainer in which there are openings of definite size all over the surface, which retains contaminants in the fuel and permits supply of dust free fuel to the fuel injection system.

FUEL LINE – Thick walled high pressure tubes connecting the delivery end of the fuel injection pump with an injector nozzle located in each of the cylinder head.

FUEL METERING – Measuring and delivering the required amount of fuel for each cycle in accordance with the engine load and delivering the same amount of fuel to each cylinder for each power stroke of the engine.

FUEL PUMP DELIVERY VALVE – assembly relieves the high pressure pipe and maintains a residual pressure in the high pressure pipe. It also reduces the work per cycle in pressurizing the fuel in the system.

FUEL TRANSFER PUMP – A mechanical device used to transfer fuel from the low level fuel tank to the injection pump.

GLOW PLUG – is an electrical heater, which is switched on for quick starting in cold weather. Its heating element protrudes directly into the combustion chamber in DI engines.

GOVERNOR – Mechanism connected to the control rack of the fuel injection pump and thus acts as an intermediate mechanism between the control rack and accelerator pedal.

HOT SMOKE – The smoke that consists of unburned carbon particles (0.5 to 1.0 micron in diameter) and other solid combustion products. This may be light gray to black smoke.

IGNITION LAG – The time interval between the start of injection and start of ignition. Also called DELAY PERIOD.

INDIRECT INJECTION ENGINES – have the combustion space divided into two parts connected by a nozzle or one or more number of orifices. The fuel is injected into the auxiliary chamber. The main chamber is situated in the piston top.

INJECTION LAG – Time interval between the closure of the fuel pump ports by the plunger and the begining of fuel injection by the injector.

INJECTION PUMP – A high variable pressure pump delivering fuel into the combustion chamber.

INJECTION SYSTEM – The components necessary for delivering fuel to the combustion chamber in correct quantity, at the correct time and in a condition satisfactory for efficient combustion.

INJECTION TIMING – Crank angle with respect to TDC at which fuel injection starts, during the compression stroke.

INJECTOR – A device for injecting fuel oil into the combustion chamber of an engine against the pressure of air within the chamber. Also called INJECTION NOZZLE.

MASKED VALVE – An inlet valve which has a mask *i.e.*, a projection, at the rear of the valve head over a part of the periphery so as to admit air in the desired direction.

MAXIMUM SPEED GOVERNOR – limits the maximum speed of operation of a diesel engine.

MECHANICAL INJECTION – Mechanical force pressurizing the metered fuel and causing injection.

METERING FUEL PUMP – A fuel pump delivering a controlled amount of fuel per cycle.

MINIMUM AND MAXIMUM SPEED GOVERNORS – are used predominantly in motor vehicles. The idle and maximum speeds are governed, but not the range in between where the quantity of fuel injected is controlled by the accelerator pedal.

MIXED CYCLE – An engine in which fuel burns partly at constant volume and partly at constant pressure. Sometimes applied to the actual combustion cycle in most high speed diesel engines.

NOZZLE – The component containing the fuel valve and having one or more orifices through which fuel is injected.

OPEN COMBUSTION CHAMBER – A type of diesel combustion chamber in which all the air meant for combustion is confined in one space and combustion of entire fuel takes place within this space. Also called DIRECT INJECTION CHAMBER or QUIESCENT CHAMBER.

OPEN TYPE NOZZLE – A simple spray nozzle open to the cylinder but with a check valve which prevents the high pressure gases in the engine cylinder from passing to the pump.

OPTIMUM INJECTION ADVANCE – Fuel injection timing before TDC which will result in minimum ignition delay.

ORDERLY TURBULENCE – Air motion which is controlled as to direction and velocity.

PENETRATION – The distance through which fuel particles are carried by the kinetic energy imparted to them when they leave the fuel nozzle.

PHYSICAL DELAY PERIOD – The time that elapses between the begining of fuel injection and the begining of preflame reactions.

PILOT INJECTION – is the early injection of a small quantity of fuel to initiate combustion of the injected main fuel, in a diesel engine.

PINTLE TYPE NOZZLE – A closed type fuel nozzle having a projection on the end of the fuel valve which extends into the orifice when the valve is closed.

PRECOMBUSTION CHAMBER – Part of the combustion space located in the cylinder head, into which fuel is injected, and combustion starts here and spreads into the main chamber via the interconnecting orifices.

QUALITY GOVERNING – Power developed is governed by varying the quantity of fuel injected while the quantity of air sucked in is almost the same. The air fuel ratio in the combustion chamber is different at different loads.

RATE OF FUEL INJECTION – Amount of fuel that is injected into the combustion chamber in unit time or in one degree of crank travel.

REENTRANT COMBUSTION CHAMBER – is an open combustion chamber which has a smaller diameter (opening) at the entry than at the middle.

RESIDUAL PRESSURE – The pressure at which the fuel is retained in the fuel line when the injector needle valve and the pump delivery valve are in the closed position.

SAC VOLUME – is the dead volume between the nozzle seat and the end of the spray holes, in a multi hole injector.

SEMI DIESEL – A diesel which utilizes injection of fuel, but also uses electric spark ignition.

SMOKE – is nothing but the carbon particles suspended in the exhaust gases. It may be blue smoke, black smoke and white smoke.

SODIUM COOLED VALVE – A valve designed to allow the stem and head to be made hollow and partially filled with metallic sodium for better cooling.

SOLID INJECTION SYSTEM – The system which injects only the metered quantity of fuel by means of a pumping device. Also called AIRLESS INJECTION SYSTEM.

SQUISH – Radial inward flow of air from the annular space above the piston; into the combustion chamber cavity during the compression stroke of the piston.

SQUISH AREA – The area confined to the cylinder head and flat surface of the piston when on compression stroke, which causes squish.

SQUISH HEIGHT – refers to the clearance between the piston top and cylinder head, at TDC. This is reduced to a minimum consistent with manufacturing capabilities and operational aspects.

SUCTION SWIRL – Rotary motion imparted to the air during suction by admitting air into the engine cylinder in a tangential direction.

SUPER CHARGING – Process of admitting into the engine cylinder, a charge larger in quantity (*i.e.*, weight) than what the cylinder would contain due to regular suction stroke *i.e.*, by natural aspiration.

SURFACE FILTER – Fuel filter similar to a sieve, consists of a number of discrete holes and pores, through a single layer of material.

SWIRL – Rotation of mass of air as it enters the engine cylinder. This is one form of turbulence.

SWIRL COMBUSTION CHAMBER – Part of the combustion space, spherical or cylindrical in shape, located in the cylinder head, in which vigorous swirl is created and into which fuel is injected and combustion starts and spreads into the space above the piston.

SWIRL RATIO – is the ratio of (air) swirl speed in the combustion chamber to engine speed.

TIMING MARK (**injection**) – The mark made on the vibration damper or flywheel, used to check injection timing.

TURBULENCE – Violent swirling motion. Fuel injection produces some turbulence. Additional turbulence is provided by the design features of the combustion chamber.

TURBULENCE CHAMBER – A combustion chamber connected to the cylinder through a throat. Fuel is injected across the chamber and turbulence is produced in the chamber by the air entering during compression.

TWO STAGE COMBUSTION – Combustion occurring in two distinct steps such as in a precombustion chamber.

UNCONTROLLED COMBUSTION – is the instantaneous combustion of the fuel that got accumulated in the combustion chamber, at the end of the delay period.

UNIT INJECTOR – A combined fuel injection pump and fuel nozzle.

VALVE CLOSING PRESSURE – is the fuel pressure at which the fuel injector needle valve snaps back on its seat. For the differential valve stem, this is less than the nozzle opening pressure.

VALVE CLOSING ORIFICE NOZZLE – In this nozzle, the needle valve sits directly on the top of the orifice. This prevents any fuel in the sac volume escaping into the cylinder after the valve is closed.

VALVE OPENING PRESSURE – is the fuel pressure at which the needle valve of the injector lifts off its seat.

VARIABLE INJECTION TIMING – Altering the injection timing as the engine speed changes (advancing timing as speed increases), by a set of flywheel weights mounted on the injector pump camshaft, to obtain better performance.

VARIABLE SPEED GOVERNOR – Governs both the idle and maximum speeds as well as the range in between.

CHAPTER 23

TWO WHEELERS

ANTI DIVE SYSTEM – A system installed with some brakes that uses front fork damping and the front brake assembly to prevent excessive fork compression and to improve handling when brakes are applied.

AUTOMATIC TRANSMISSION – A transmission not using a manually operated clutch.

AXLE – A shaft used to support a part or parts across the frame or forks. *e.g.*, front and rear axles.

BACK BONE FRAME – Frame which uses the engine as a structural member for load carrying.

BEAD – The portion of the tyre which holds it onto the rim.

BEZEL – Piece of metal surrounding head lights, gauges or similar components, sometimes used to hold the glass face of a gauge in the dashboard.

BOOSTER PORT – In a two stroke engine, the port that allows an extra amount of air fuel mixture from the intake port into the combustion chamber.

BRAKE ACTUATOR CAM – Small cam that pivots in brake backing plate and forces brake shoes into brake drum.

BRAKE CALIPER – Part of a disc brake which holds friction pads and encloses disc. As the brake is applied, hydraulic fluid forces a piston in caliper towards disc, causing disc to be pinched between brake pads.

BRAKE DISC – A round, flat disc made of steel or cast iron. It is mounted on outside of wheel hub.

BRAKE DRUM – A circular ring of cast iron that is part of wheel hub. It provides a place for brake lining to be applied.

BRAKE FADE – Loss of braking power, usually caused by excessive heat after repeated brake applications.

BRAKE LINE – Special hydraulic tubing made of steel, plastic or reinforced rubber. Hydraulic lines must be capable of withstanding extreme pressure without deforming.

BRAKE LINING – A special high friction material made of asbestos and other materials bonded to brake shoes and brake pad plates. Brake lining produces friction and heat when it is forced against brake drum.

BRAKE PAD – The friction pad on a disc brake system.

BRAKE SHOE – The friction lining on a drum brake system.

BRAKE WEAR INDICATOR – Index grooves, tabs, or reference lines to indicate amount of brake lining or pad wear.

BRIDGED PORTS – A vertical port division in a two stroke cycle engine cylinder which allows use of a large port without the danger of ring or piston catching.

CAST ALLOY WHEEL – A one piece wheel made of cast aluminium or magnesium alloy. This design is more rigid than the wire spoked wheel.

CENTRIFUGAL CLUTCH – Clutch engaged by centrifugal force as engine speeds up.

CHAIN DRIVE – Use of a chain and sprockets to connect gear box output shaft to rear wheel.

CHAIN STRETCH – Wear of pins and bushings of a roller or hyvo chain, causing the chain to lengthen.

CHAMFER – To bevel an edge of an object or to chamfer edges of port openings in a two stroke cycle cylinder to prevent piston ring breakage.

CLOSE RATIO GEAR BOX – A gearbox with gear ratios spaced close together.

CLUTCH – Device used to connect and disconnect engine power to gearbox input shaft.

CLUTCH BASKET – Part of clutch assembly containing drive plates. Primary drive gear engages teeth on the outside of the clutch basket.

CLUTCH HUB – Part of the clutch that engages with plain driven clutch plates. Clutch hub is mounted on the gearbox input shaft.

CLUTCH PRESSURE PLATE – Part of a clutch assembly providing pressure against the clutch disc or clutch plates.

CLUTCH RELEASE MECHANISM – Mechanism that moves the clutch pressure plate away from the clutch pack allowing the clutch to slip.

CONICAL HUB – A wheel hub (wire wheel) with spoke holes on the brake side of a wheel at a greater distance from the center of the hub than spoke holes on the opposite side of the hub.

CONVENTIONAL REAR SUSPENSION – Suspension used on dual purpose and road bikes which provides less than 152 mm of suspension travel.

COUNTERSHAFT SPROCKET – Output sprocket from gearbox. Mounted on output shaft in indirect drive gearbox and on high gear pinion in direct drive gear box.

CRADLE FRAME – Frame built of tubing which supports and surrounds the engine.

CRANKSHAFT AXLES – Extension at each end of the crankshaft to provide a mounting place for the main bearings, primary drive gear or sprocket, and alternator rotor or magneto flywheel.

CURB WEIGHT – The weight of a vehicle without passengers or payload, but including all fluids (oil, gas, coolant etc.) and other equipment specified as standard.

DAMPER – Device which uses oil metered through orifices to control abrupt suspension movement during expansion and compression.

DAMPER ROD – Tube secured to the bottom of each fork slider to hold the slider onto the fork leg. Damper rod controls movement of front suspension by metering hydraulic fluid through orifices in the rod.

DIAPHRAGM SPRING – A slightly cone shaped metal disc which acts as a clutch pressure plate spring when flattened.

DIRECT DRIVE GEARBOX – Power is transmitted from clutch to input shaft (main shaft), to layshaft, to high gear pinion, which has output sprocket mounted on it.

DISC BRAKE – A brake consisting of a flat circular disc attached to the wheel. A hydraulic or mechanical caliper applies pressure to two brake pads and press them on the disc to slow or stop disc rotation.

DOUBLE PISTON CALIPER – A hydraulic brake caliper with two pistons and provisions for applying hydraulic pressure equally to both pistons. The caliper body is fixed solidly.

DOUBLE ROW CHAIN – A chain having two rows of rollers. Duplex (double row) chains are used for primary drives.

DRIVE PLATE – A clutch plate which is indexed into the clutch basket (outer hub) by tabs. Drive plate has friction material bonded to its surface. When clutch is engaged, drive plate transfers power to driven plate.

DRIVEN PLATE – A clutch plate which is indexed onto the clutch inner hub by tabs or splines around its inside diameter. Driven plate is usually a plain plate (no friction material) and drives gearbox input shaft through clutch inner hub.

DRIVE TRAIN – The drive system that transfers the power produced by the engine to the rear wheel, includes primary drive, clutch, transmission and fluid drive systems.

DRUM BRAKE – A brake consisting of two brake shoes mounted on a backing plate. One or two cams cause the shoes to expand against inside of the brake drum. Brake drum is part of hub or is bolted to hub.

DRY CLUTCH – A clutch assembly that does not run in an oil bath.

ELLIPTICAL PORT SHAPE – Rounded port shape designed to prevent ring catching in large ports of two stroke cycle engines.

ENDLESS CHAIN – A roller chain without a master link for connection of ends. All pin links are permanently riveted.

EXHAUST PORT – In a two stroke engine, the passage way located in the upper portion of the cylinder opposite from the intake port, that channels spent combustion gases to the exhaust system.

EXHAUST PORT TIMING – Amount of time, two stroke cycle exhaust port is open, expressed in crankshaft degrees or piston travel.

EXTENSION – The return or stretching outward of suspension components (after compression) caused by spring pressure.

FINAL DRIVE RATIO – The number of times the transmission output shaft turns to produce one revolution of the rear wheel.

FINAL DRIVE SYSTEM – The part of the drive train that uses a chain, belt or shaft to direct the power flow from the transmission to the rear wheel.

FORK SLIDERS – Lower portion of fork which slides over the fork leg.

FORK TUBES – Long sturdy tubes attached to triple clamps and fitted inside fork sliders.

FRAME – The skeleton of the two wheeler made of tubes, steel plates or pressed steel that supports the rider and engine and provides attachment points for the frame components.

FRONT FORK – The spring and damping device that holds the front wheel in place.

FUEL PET COCK – An on-off valve located at the bottom of fuel tank. It may provide for reserve fuel supply and may have a filter screen and sediment bowl. Fuel line to carburettor is attached to fuel petcock.

GAS CHARGED SHOCK ABSORBER – A shock absorber using a pressurized gas such as nitrogen to help prevent changes in damping as shock absorber heats up.

GUSSET – A reinforcing plate or boxed section used to prevent flexing of frame or swing arm.

HUB – The center part of a wheel.

HYDRAULIC DAMPING – A conversion of oil resistance to heat energy to create a force that opposes input motion.

HYDRAULIC DISC BRAKE – A brake system that exerts pressure through brake fluid to press brake pads against a disc to slow or stop the vehicle.

HYDRO-PLANING – A phenomenon of driving when water builds up under the tyre tread, causing it to lose contact with the road. Slowing down will usually restore normal tyre contact with the road.

HY-VO-CHAIN – A very strong chain made of toothed plates positioned side by side and held together by pins. Advantage of this chain is great strength and quiet operation.

INDIRECT DRIVE GEARBOX – A gearbox where power is transferred from clutch to input shaft and from that to output shaft. Output shaft has a sprocket mounted on one end.

INPUT SHAFT – Shaft of a gearbox which carries power into gearbox. Clutch assembly is mounted on gearbox input shaft.

INTAKE PORT – In a two stroke engine, the passage way that allows the air fuel oil mixture to travel from the carburettor into the crankcase, located on the lower half of the cylinder opposite from the exhaust port on all but rotary valve engine.

INTAKE TIMING – Amount of time, two stroke cycle engine intake port is open, expressed in crankshaft degrees or piston position.

INTERNAL COOLING – Engine cooling provided by oil, fuel mixtures, and valve overlap.

LACING – The positioning of spokes in the hub and rim of a wheel.

LATERAL RUNOUT – Side to side movement (wobble) of a wheel rim.

LAY SHAFT – Second shaft in a direct drive gearbox. It transmits power from input shaft to high gear pinion.

LOAD RANGE – Indicates the number of plies at which a tyre is rated, load range B equals 4 ply rating, C equals 6 ply rating, and D equals 8 ply rating.

LONG TRAVEL SUSPENSION – Suspension used on current competition dirt bikes to provide more than 150 mm of travel.

MASTER LINK – A pin link which has one removable side plate located by a clip. This allows a convenient way of separating the chain.

MECHANICAL BRAKE – A brake system which uses a mechanical advantage by way of levers and cables or rods to apply brakes. A brake system not using hydraulic fluids or hydraulics.

MILD PORT TIMING – Two stroke cycle engine ports open for a relatively short time, providing for a broad power band.

MOTOR CYCLE – A two wheeled vehicle powered by an internal combustion engine.

MULTI PLATE CLUTCH – A clutch assembly using more than one driving plate and more than one driven plate.

MULTIPLE PORTS – Use of many small transfer ports rather than two large ports in the two stroke cycle engine cylinder.

NON-PRIMARY KICKSTART – A kickstart system which uses gearbox input shaft and clutch hub to connect kickstart lever to crank shaft. For starting , gearbox must be in neutral and clutch engaged.

NON-UNIT CONSTRUCTION – Engine design with separate engine crankcase and gearbox case.

O-RING CHAIN – Roller chain which uses 'O' rings to permanently seal lubricant into area between pins and bushings.

OUTPUT SHAFT – Gearbox shaft that transmits power to final drive at a selected ratio.

OVERALL GEAR RATIO – Ratio of crankshaft revolutions to rear wheel revolutions.

PISTON (brake system) – A movable part of a master cylinder or wheel cylinder. Master cylinder piston produces pressure and caliper (wheel cylinder) piston applies that pressure to brake disc, slowing or stopping the wheel rotation.

PISTON PORT SYSTEM – The two stroke intake system that uses the piston to open and close the intake and exhaust ports.

PLY RATING – A rating given to a tyre which indicates strength (but not necessarily actual number of plies).

PORTS – The passage ways cut into a two stroke engine cylinder, that channel gases into and out of the engine. Also called PORT WINDOWS.

POWER TO WEIGHT RATIO – Ratio of horse power to weight of a vehicle.

POWER TRANSMISSION – A system of gears, chains, sprockets and shafts that transfer power from the crankshaft to rear wheel in varying ratios.

PREMIX – Two stroke cycle engine lubrication provided by oil mixed with fuel in fuel tank.

PRIMARY CHAIN DRIVE – A primary drive design that uses a chain and sprockets to transmit engine power from crankshaft to clutch.

PRIMARY DRIVE SYSTEM – A system of gears, chains or belts that transmits power from the crankshaft to the transmission, includes the clutch.

PRIMARY KICKSTART – A kickstart system which connects kickstart lever to crankshaft through clutch basket. Gearbox can be in gear for starting if clutch is disengaged.

PRIMARY REDUCTION – A gear reduction usually about three to one, accomplished by primary drive.

RADICAL PORT TIMING – Ports open for a relatively long period of time.

REED CAGE – An aluminium frame providing a mounting place for reed petals in a two stroke cycle engine.

REED PETALS – Movable part of a reed valve assembly.

REED VALVE SYSTEM – A two stroke intake system that uses thin flapper valves that are opened by vacuum and closed by pressure.

REAR SUSPENSION – The suspension system that consists of the swing arm, rear shock absorbers, springs and linkage.

RIM BAND – A strip of soft rubber that protects the inner tube from the spoke nipples.

RIM OFFSET – The deliberate lacing of spokes off center to the hub to achieve rear chain or front brake clearance.

ROLLER CHAIN – A chain made up of pins, side plates, bushings and rollers. Roller links are connected by pin links to achieve desired length. The ends are usually connected by a pin link which has a removable side plate, called master link.

ROTARY VALVE – A two stroke engine intake system that uses a partially cut away rotating disc to open and close the intake passage into the crankcase.

SCAVENGING – Inertia caused movement of gases through cylinder and exhaust system of a two stroke cycle or four stroke cycle engine by the incoming fresh charge.

SEALED BEAM – A modern automotive headlight. The lens, reflector and filament form a single unit.

SEDIMENT BOWL – A cup located at the bottom of the fuel petcock, designed to prevent flow of dirt and water into the fuel line.

SHAFT DRIVE – Use of a drive shaft, universal joints, and gears to connect gearbox output shaft to rear wheel.

SHOCK ABSORBER – Rear suspension unit made up of a coil spring and a hydraulic damper. It prevents spring oscillations.

SINGLE ROW CHAIN – A conventional roller chain with one row of rollers.

SOLID ROLLER CHAIN – Roller chain using a one piece roller and bushing assembly.

SPINDLE (**steering fork**) – A shaft which is secured to upper and lower tripple clamps. It pivots on bearings mounted in the steering head.

SPOKE – A metal part, often a wire, that runs between the rim and hub of a spoked wheel to transfer force to the hub.

SPOKE CROSSING PATTERN – Number of spokes crossed on the same side of the hub by anyone spoke.

SPOKED WHEEL – A wheel consisting of a rim, spokes, nipples and hub. Spokes are laced between hub and rim and are attached to the rim by nipples.

SPOKE TORQUE WRENCH – A small torque wrench used to adjust accurately the spoke tension.

SPONGY LEVER or PEDAL – A soft or spongy feeling when the brake lever or pedal is depressed. It is usually due to air in the brake lines.

SPROCKET – A circular plate with teeth machined around outside to engage links of a chain.

SPRUNG WEIGHT – The weight of the two wheeler components that are supported on springs, includes the fork tube, engine, tripple clamp and frame.

STAMPED FRAME – A frame stamped from pieces of sheet metal which are welded together to provide support for engine and suspension.

STAMPED WHEEL – A wheel assembly using stamped sheet metal spokes in place of small wire type spokes. A stamped wheel resembles a cast alloy wheel in appearance.

STEERING DAMPER – A device which uses friction or a hydraulic damper to reduce steering oscillation.

STEERING HEAD – The tube at the top and front of the frame that supports the steering stem and is welded at a specific angle to provide the proper fork angle.

STEERING OFFSET – Distance between steering axis and axis on which the axle pivots. Steering offset is accomplished by offsetting axle or tripple clamps.

STEERING RAKE – Angle of steering axis from vertical, given in degrees.

STEERING STEM – A shaft positioned through the steering head that enables the front end to turn.

SUSPENSION SYSTEM – A system that consists of front forks, shock absorbers, springs and the swing arm, used to support the two wheeler on its axles and wheels.

SWING ARM – The assembly that links the rear wheel to the frame and allows the rear wheel to move up and down, and prevents the wheel from moving laterally and flexing at the pivot point.

SWING ARM PIVOT – Forward part of swing arm, where it is attached to the frame and pivots on bushings or bearings.

SYMMETRICAL HUB – A wheel hub (wire wheel) which has spoke holes on each side of the hub at the same distance from hub center.

TELESCOPIC FORK – Front suspension unit made up of two fork tubes and two sliders that telescope up and down the tubes against spring pressure.

THROTTLE CABLE – A cable consisting of an outer housing and an inner cable which connects carburettor to throttle twist grip.

THROTTLE STOP SCREW – An idle speed adjustment screw used in slide type carburettor. This screw contacts base of throttle slide. As the screw is turned in, throttle slide is raised, increasing idle speed.

THROTTLE TWIST GRIP – A device mounted on the end of handle bar which locates one end of outer throttle cable and pulls inner cable as the twist grip is rotated.

TRAIL – The distance between the center of the tyre contact patch and the imaginary point where the steering head axis strikes the ground.

TRANSFER PORT – Opening in the cylinder wall of a two stroke engine which connects cylinder to crankcase.

TRANSFER PORT TIMING – Amount of time transfer port is open, expressed in crankshaft degrees or piston position.

TRANSMISSION – The part of the drive train that uses a series of gears and shafts to increase the torque and provides gear shifting to allow increases in speed without overworking the engine.

TREAD WEAR PATTERN – The pattern of wear on the tyre which can be read to diagnose problems in the front suspension.

TRIPPLE CLAMPS – An assembly that consists of the steering stem and two brackets that position the fork tubes.

TRIPPLE ROW CHAIN – A chain having three rows of rollers. Tripple row chains are used for primary drives.

TWO STROKE CYCLE – A cycle of overlapping events (intake, transfer, compression, power and, exhaust) which all occur during one revolution of crankshaft.

TYRE PRESSURE GAUGE – A pressure gauge used to check tyre air pressure.

UNIFIED BRAKE SYSTEM – A system that automatically coordinates the front and rear brake systems when the rider applies pressure to the brake pedal.

UNIT CONSTRUCTION – Engine design which has one crankcase to house gearbox, clutch, primary drive and engine.

UNIVERSAL JOINT – A flexible joint which allows changes in angle of drive shaft.

UNSPRUNG WEIGHT – The weight of the two wheeler components that are not supported on springs, includes the wheel assembly, fork sliders, rear shocks, a portion of the swing arm, the chain or shaft drive, one half of the spring and on some models, the front fender.

WET CLUTCH – A multiple plate clutch which runs in an oil bath (primary drive).

WHEEL BALANCING – An equal distribution of weight achieved by determining where the wheel is heavier and then placing weights opposite the heavy portion.

WHEEL BASE – Distance between the center of front wheel and the center of rear wheel.

WHEEL WEIGHT – Small weights attached to the wheel to balance the wheel and the tyre assembly.

WIDE RATIO GEARBOX – A gearbox having wide ratio spacing between gears. A wide ratio gear box is used in off road and trials motor cycles.

CHAPTER 24

AUTOMOTIVE VEHICLES

ACCELERATOR – A pedal connected to the carburettor throttle valve of a motor vehicle, or to the fuel injection control where oil engines are used.

ACKERMAN PRINCIPLE – Steering geometry in which the outer ends of the steering arms are bend slightly inward so that when the vehicle is making a turn, inside wheel will turn more sharply than the outer wheel. This principle produces toe out on turns.

ACKERMAN STEERING – Arrangement whereby a line extended from the track arms, when the wheels are set straight ahead, should meet on the chassis centre line at 2/3 of the wheel base from the front, allowing inner stub axle to move through a greater angle than the outer.

AERODYNAMIC DRAG – is the air resistance to the motion of the vehicle. This consists of profile drag, induced drag, skin friction drag, interference drag, and cooling and ventilation drag.

AERODYNAMIC LIFT – is the vertical component of the resultant force caused by the pressure distribution on the vehicle body.

AIR BLEEDER – A device used to remove air from a hydraulic system. Types include a needle valve, capillary tubing to the reservoir, and a bleed plug.

AIR BRAKE – A braking system which uses compressed air to supply the effort required to apply brakes.

AIRFOIL – Device, similar to a stubby wing.

AIR RESISTANCE – The motion of an automobile is associated with the displacement of air particles, which requires some power of the engine.

Air resistance depends on the size and shape of the vehicle body, speed of the vehicle and wind velocity.

AIR SPRING – Container and plunger are separated by air under pressure. When container and plunger attempt to squeeze together, air compresses and produces a spring effect. Air spring has been used in some suspension systems.

ALL WEATHER TYRE – A tyre designed to provide good traction on dry, wet and dirt and snow covered roadways.

ANTI DAZZLE MIRROR – One having a photoeltric control circuit which changes it from a fully reflecting condition to partial reflection from a glass air interface when actuated by the head lamp beam of a following vehicle.

ANTILOCK BRAKE SYSTEM (ABS) – If the brakes are applied so hard that the wheels tend to stop turning and thus a skid starts to develop, the antilock brake system comes into operation and partly releases the brakes. This makes the wheels continue to rotate. However, intermittent braking continues. But it is held below the point where the skid would start.

ANTIMIST PANEL – A panel fitted to the rear window enclosing a volume of still air between itself and the outer glass.

ANTIROLL BAR – Torsion bar mounted transversely in the chassis in such a way so as to counteract the effect of opposite spring deflections.

ARCH – The curve of a leaf spring. If the centre is lower than the ends, it is called positive arch, if the centre is higher than the ends, it is called negative arch.

ARCING (brakes) – Grinding new brake linings to the same diameter (arc) as that of the brake drum surface.

ASPECT RATIO – The ratio of the width to the length. On tyres, it is the fully inflated height divided by the cross section.

AUTOMOBILE – is a self propelled vehicle. The power required to propel the vehicle is supplied by the engine (also called prime mover). Scooters, motor cycles, cars, buses, trucks etc., are different types of automotive vehicles.

AUTOMATIC LEVEL CONTROL – A suspension system which compensates for variations in load in the rear of the car, positioning the rear at a predesigned level regardless of load.

AUTOMATIC TRANSMISSION – A power transmission system for road vehicles, in which the approximately optimum engine speed is maintained through mechanical or hydraulic speed changing devices which are automatically selected and operated by reference to the road speed of the vehicle.

AXLE – A cross bar supporting a vehicle on which one or more wheels turn.

AXLE (full floating) – Axle used to drive rear wheels. It does not hold the wheels on nor support them.

AXLE (semiquarter or one quarter floating) – Axle used to drive wheels, hold them on and support them.

AXLE THREE QUARTER FLOATING – Axle used to drive rear wheels as well as hold them on and support them.

AXLE FLANGE – A flat surface on the outboard end of the axle shaft to allow wheel attachment.

AXLE GEAR – A gear in the differential carrier that drives the driving wheels.

AXLE RATIO – Relationship or ratio between the number of times the propeller shaft or drive shaft must revolve to turn the axle shafts one turn.

AXLE SHAFT – The shaft used to transmit power from the differential to the wheels.

BACK LOCKING – The steering gear is so constructed that it is easy to turn the vehicle by steering wheel, but it is difficult to turn the steering wheel by turning the front wheels. This back locking prevents the bumps and shocks experienced by the wheel on the road surface from being transmitted to the steering wheel.

BACKING PLATE – A mounting plate that holds the brake shoes, cam lever, pivot pins and springs inside the brake drum.

BALL JOINT – Flexible joint utilizing ball and socket type of construction, used in steering linkage set ups, steering knuckle pivot supports etc.

BALL JOINT ROCKER ARMS – Rocker arms that instead of being mounted on shaft, are mounted upon ball shaped devices on end of stud.

BALL JOINT STEERING KNUCKLE – Steering knuckle that pivots on ball joints instead on king pin.

BALL JOINT SUSPENSION – A type of front suspension, which does not use a steering knuckle. Instead, the wheel spindle is attached directly to the upper and lower suspension arms through ball joints. Allows movement up and down as well as rotation.

BALL STUD – Stud with a ball on end, commonly used in steering linkage to connect pitman arm to linkage, or to connect tie rods.

BALL AND TRUNNION JOINT – A type of universal joint which combines the universal joint and slip joint in one assembly.

BEAD (tyre) – Steel wire reinforced portion around a tyre opening that engages the wheel rim.

BELL HOUSING (clutch housing) – Metal (cast iron or aluminium) cover that surrounds flywheel and clutch, or torque converter assembly.

BELL MOUTH – The taper of a brake drum.

BELTED TYRE – A tyre that is reinforced with a build up of cord under the tread area.

BENCH BLEEDING – Process of removing air from the master cylinder pressure area before installing it in the vehicle.

BENDIX TYPE STARTER – A self engaging starter drive gear. Gear moves into engagement when starter armature shaft starts spinning and automatically disengages when starter stops and engine speed increases.

BIAS BELTED TYRE – A tyre in which plies are laid on the bias, criss crossing each other, with a circumferential belt on top of them. The rubber tread is vulcanized on top of the belt and plies.

BINDERS – Compounds that hold the friction materials together in brake linings.

BLEEDING – Removing air, pressure, fluid etc. from a closed system as in the brake system or air conditioning system.

BLEEDING (brakes) – Removal of air from hydraulic system. Bleeder screws are loosened at each wheel cylinder (one at a time) and brake fluid is forced from master cylinder through lines until all air is expelled.

BLEEDING (steering) – A process by which air is removed from a hydraulic system (power steering) by bleeding off part of the fluid or operating the system to work without the air.

BODY – The assembly of sheet metal sections together with windows, doors, seats and other parts, that provides an enclosure for the passengers, engine and so on.

BODY PANELS – Sheets or panels of steel which are fastened together by welding to form the vehicle body.

BODY ROLL – The vehicle body leaning sideways as the vehicle turns.

BOGIE – A small truck, of short wheel base running on rails, commonly used for the conveyance of coal, gold or other ores, concrete etc.

BONDED BRAKE LINING – Brake lining that is attached to the brake shoe by adhesive.

BONNET – British term for car hood.

BOOSTER – Device incorporated in a car system (such as brake and steering), to increase pressure output or decrease amount of effort required to operate or both.

BORG WARNER OVER DRIVE – A method of reducing engine rpm in relation to road speed. The unit is attached at the rear of the gear box and operates through epicyclic gears.

BRAKE – An energy conversion device that converts the energy of motion into heat energy and thereby slows down or stops a moving vehicle.

BRAKE (disc type) – Braking system which uses steel disc with caliper type lining application. When brakes are applied, section of lining on the caliper piston on each side of the spinning disc is forced against the disc thus imparting braking force. This type of brake is very resistance to brake fade. Also called disc brake system.

BRAKE ANCHOR – Steel stud upon which one end of brake shoes is either attached to or rests against. Anchor is firmly affixed to backing plate.

BRAKE ANTIROLL DEVICE – Unit installed in brake system to hold brake line pressure when car is stopped on upgrade, and brake pedal is released. Antiroll device will keep brakes applied until either clutch is released or, as in some models, accelerator is depressed.

BRAKE BACKING PLATE – Rigid steel plate upon which brake shoes are attached. Braking force applied to shoes is absorbed by backing plate.

BRAKE BAND – Band faced with brake lining, that encircles a brake drum. Used on several parking brake installations.

BRAKE BIAS – The stopping effort of the front wheels compared to that of the rear wheels.

BRAKE CALIPER – The hydraulic cylinder at the wheel used to apply the disc brake linings against the rotor.

BRAKE CLEARANCE – is the clearance provided between the lining and the drum or disc. Wear and tear of the lining increases this clearance and hence to be adjusted periodically.

BRAKE DRUM – Metal drum mounted to the vehicle wheel which forms the outer shell of the brake. Brake shoes when moved out or moved apart press against the rotating drum to slow or stop drum and wheel rotation.

BRAKE EFFECTIVENESS – is how effectively the brakes perform their function. This depends on the area of the brake lining, amount of pressure applied to the brake shoes, radius of the brake drum, vehicle wheel radius, coefficient of friction of braking surfaces and coefficient of friction between the tyre and the road surface.

BRAKE FADE – A reduction or fading out of braking effectiveness due to loss of friction between brake shoes and drum. This is caused by overheating (heat build up) from excessively long and hard brake application for instance, when coming down a long hill or mountain.

BRAKE FEEL – The reaction of the brake pedal against the drivers foot, that tells him how heavily he is applying the brakes.

BRAKE FLUID – A special non -mineral oil fluid used in hydraulic braking system. Never use anything else in place of regular fluid.

BRAKE FLUSHING – Cleaning brake system by flushing with alcohol or brake fluid. Done to remove water, dirt or any other contaminant. Flushing fluid is placed in master cylinder and forced through lines and wheel cylinders where it exits at cylinder bleed screws.

BRAKE LINE – Special hydraulic tube made of steel, plastic or reinforced rubber suitably designed to withstand extreme pressure without deforming.

BRAKE LINING – A special high friction material made of asbestos and other materials bonded to brake shoes and brake pad plates. Brake lining produces friction and heat when it is forced against the brake drum or disc.

BRAKE PULL – A condition in which the vehicle turns each time the brakes are applied.

BRAKE ROTOR – The brake friction surface that rotates at wheel speed designed for contact with the brake pads on disc brake assemblies.

BRAKE SELF ADJUSTERS – A cable operated device used to adjust brake shoes automatically.

BRAKE SHOES (disc brakes) – Flat metal pieces lined with brake lining which are forced against the rotor face. Also called brake pads.

BRAKE SHOES (drum brakes) – Arc shaped metal pieces lined with heat resistant fibre. When forced against the brake drum, stops wheel rotation.

BRAKE SHOE HEEL – End of brake shoe adjacent to anchor bolt or pin.

BRAKE SHOE TOE – Free end of brake shoe, not attached to or resting against an anchor pin.

BRAKING SYSTEM EFFICIENCY – is measured in terms of the rate at which brake will bring the vehicle to a stationary position from a given speed. It is expressed as the ratio of the vehicle deceleration rate to the acceleration due to gravity.

BREAKE (tyre) – Rubber or fabric (or both) strip placed under the tread to provide additional protection for main tyre carcass.

BULK HEAD – The structural part of the vehicle connecting the front of the floor assembly to the roof structure.

BUMPER – which is attached to the vehicle frame takes the shock of impact or collision and transfer the same to the frame. By this means, damage to engine parts, radiator, lamps etc. is avoided.

BUMP STEER – The steering effect caused by the suspension moving through its travel.

CALIPER – A housing for the hydraulic components of a disc brake system.

CAMBER – Tilting of the top of wheels from the vertical, when the tilt is outward, camber is positive.

CAMBER ANGLE – The outward (positive) or inward (negative) angle of the wheel centre line to absolute vertical.

CARDAN UNIVERSAL JOINT – A universal joint of the ball and socket type.

CARLIFT – An air, electrical or hydraulically operated piece of shop equipment which can lift the entire vehicle, or in some cases, one end of the vehicle.

CARRIER BEARINGS – Bearings upon which differential case is mounted.

CASING OF TYRE – The tyre casing, made of fabric or cord to which rubber is vulcanized. It is the outer part of the tyre assembly.

CASTER – The tendency of a wheel to follow the direction of the pivot movement. Tilt of the top of the king pin forward or backward from the vertical. When tipped forward it is called negative caster. Backward tilt from the vertical is called positive caster.

CASTER ANGLE – The rearward (positive) or forward (negative) angle of the steering axis to absolute vertical.

CENTRE STEERING LINKAGE – Steering system utilizing two tie rods connected to steering arms and to central idler arm. Idler arm is operated by drag link that connects idler arm to pitman arm.

CENTRIFUGAL CLUTCH – Clutch that utilizes centrifugal force to expand a friction device on driving shaft until it is locked to a drum on driven shaft. The clutch comes into action as it spins faster.

CHANNELED – Car body lowered down around frame.

CHASSIS – Generally chassis refers to the unit that consists of frame, engine, front and rear axles, springs, steering and brake systems, controls, drive train and fuel tank. It is an assembly of mechanisms that make up the major operating part of the vehicle. In short, it is assumed to include everything except the vehicle body and fenders.

CLASSIC or NORMAL CONTROL TRUCK – has the engine located in front of the driver's cabin.

CLUTCH – Device used to connect or disconnect flow of power from one unit to another. In a vehicle, the mechanism in the power train that connects the engine crankshaft to or disconnects it from the transmission and thus with the remainder of the power train.

CLUTCH CHATTER – A shaking or shuddering of the vehicle as the clutch is operated.

CLUTCH DIAPHRAGM SPRING – Round dish shaped piece of flat spring steel, used to force pressure plate against clutch disc in some clutches.

CLUTCH DISC – Part of the clutch assembly splined to transmission clutch or input shaft, faced with friction material. When clutch is engaged, disc is squeezed between flywheel and clutch pressure plate.

CLUTCH DRAG – A problem in which the clutch disc does not come to a complete stop after the clutch pedal is depressed.

CLUTCH EXPLOSION – Clutches have literally flown apart (exploded) when subjected to high rotational speed. Scatter shield is used on competition cars to protect driver and spectators from flying parts in event clutch explodes.

CLUTCH HOUSING – A metal housing that surrounds the flywheel and clutch assembly.

CLUTCH LINKAGE – The rods and levers that allow the driver to operate the clutch.

CLUTCH PEDAL – A pedal in the drivers compartment that operates the clutch.

CLUTCH PEDAL FREE TRAVEL – Specified distance clutch pedal may be depressed before throw out bearing actually contacts clutch release fingers.

CLUTCH PILOT BEARING – A small bronze bushing or ball bearing positioned in the crankshaft end or centre of flywheel, used to support outboard end of transmission input shaft.

CLUTCH PRESSURE PLATE – Part of a clutch assembly, that through spring pressure, squeezes clutch disc against flywheel thereby transmitting driving force through the assembly. To disengage clutch, pressure plate is drawn away from the flywheel via linkages.

CLUTCH SEMI CENTRIFUGAL RELEASE FINGERS – Clutch release fingers that have a weight attached to them, so that at high rpm release fingers place additional pressure on clutch pressure plate.

CLUTCH SHAFT – The shaft on which the clutch is assembled, with the gear that drives the countershaft in the transmission on one end. It has external splines that can be used by a sychronizer drum to lock the clutch shaft to the main shaft for direct drive.

CLUTCH SLIPPAGE – A condition in which the engine overrevs during shifting or acceleration.

CLUTCH THROWOUT FORK – In the clutch, a Y shaped member into which is assembled the throwout bearing.

CLUSTER or COUNTER GEAR – Cluster of gears that are all cut on one long gear blank. Cluster gears ride in the bottom of transmission. Cluster provides a connection between transmission input shaft and output shaft.

COEFFICIENT OF FRICTION – An index of the frictional characteristics of a material.

COEFFICIENT OF ROLLING RESISTANCE – is numerically equal to the ratio of the force causing uniform rolling of the wheel to the normal reaction of the road.

COIL SPRING CLUTCH – A clutch using coil springs to hold the pressure plate against the friction disc.

COLLAPSIBLE STEERING COLUMN – is the steering column which will collapse in its length due to the impact of the driver on to the steering wheel, on a front end crash. This safety device prevents the possibility of the driver getting injured.

CONE CLUTCH – Clutch utilizing cone shaped member that is forced into a cone shaped depression in the flywheel, or the other driving unit. Although no longer used in cars, cone clutch finds some applications in small riding tractors, heavy power movers etc.

CONSTANT VELOCITY JOINT – Two closely coupled universal joints arranged so that their acceleration and deceleration effects cancel out each other, resulting in an output driven shaft speed to be always identical with drive shaft speed.

CONTACT PATCH – The part of a tyre that is in contact with the road surface.

CONTROL ARM – A suspension member mounted horizontally with one end attached to the frame and the other end the knuckle or axle housing.

CORD – A string or thread that makes up the fabric used in tyreplys.

CORNERING WEAR – A type of tyre tread wear caused by taking turns at excessive speeds.

COUNTERSHAFT – Intermediate shaft that receives motion from one shaft and transmits it to another. It may be fixed (gears turn on it) or it may be free to rotate. In the transmission countershaft is driven by the clutch gear, gears on the countershaft drive gears on the main shaft when the latter are shifted into gear.

COWL – Part of car body between engine firewall and front of dashboard.

CROSS SHAFT (steering) – Shaft in steering box that engages steering shaft worm. Cross shaft is splined to pitman arm.

CURB WEIGHT – The weight of the complete vehicle with its normal load, less driver and passengers but with a full tank of fuel, engine and vehicle oil and coolant.

CUT OUT – operates as an automatic switch which connects and disconnects the battery with the generator, according to the speed of the latter.

DAMPERS – are nothing but a piston in a cylinder filled with oil or gas. The damper restrains undesirable bounce of the sprung vehicle mass and restrains the wheel assembly from loosing ground contact by being excited at its natural frequency.

DASH BOARD – Part of body containing driving and control instruments, switches etc.

DEAD AXLE – Axle that does not rotate or deliver power but merely forms a base upon which wheels may be mounted.

DEDION – Rear axle set up in which driving wheels are attached to the frame by a central pivot. Differential unit is bolted to frame and is connected to the driving wheels by drive axles.

DEPENDENT SUSPENSION – Wheel connected through an axle member so that movement of one wheel moves the other wheel.

DIAGONAL SPLIT BRAKE SYSTEM – A brake system design that will allow application of brakes on one front wheel and a diagonally opposite rear wheel, when part of brake system fails.

DIAPHRAGM CLUTCH – Uses a diaphragm or conical spring instead of coil springs to produce adequate pressure required for keeping the clutch in the engaged position.

DIFFERENTIAL – A mechanism between axles that permit one wheel to turn at a different speed than the other while transmitting power from the drive shaft to the wheel axles, when the vehicle is negotiating a turn.

DIFFERENTIAL CASE – A steel unit to which the ring gear is attached. Differential case drives spider gears and forms an inner boaring surface for axle and gears.

DIFFERENTIAL LOCK – The differential lock grips one or both of the side gears to the differential case. This prevents their rotation on the pins. This enables a larger torque to be transmitted to the gripping wheel than that to the slipping wheel.

DIRECT ACTING SHOCK ABSORBER – Type of shock absorber which shortens or lengthens in action. Also called telescopic shock absorber.

DIRECT DRIVE – Such as high gear when crankshaft and drive shaft revolve at same speed.

DIRECTIONAL STABILITY (steering) – Ability of vehicle to move forward in straight line with minimum of driver control. Vehicle with good directional stability will not be unduly affected by side wind, road irregularities etc.

DISC BRAKE – When the brake pedal is depressed, pads lined with friction material are forced towards one another. In doing so, they come in contact with the disc (attached to the wheel) which normally rotates between them. This provides braking effort.

DISC WHEEL – Wheel constructed of stamped sheet.

DIVE – The front wheel of the vehicle lowering during braking.

DOLLY BLOCKS – Blocks of metal, variously shaped and contoured, used to straighten body panels and fenders. The dolly block is held on one side of the panel while the other side is struck with a special hammer.

DOUBLE LEADING SHOE – A drum brake having two leading shoes and no trailing shoes. Each shoe has its own actuating mechanism and pivot.

DOUBLE PISTON CALIPER – A hydraulic brake caliper with two pistons and provision for applying hydraulic pressure equally to both pistons. The caliper body is fixed solidly.

DOWN SHIFT – Shifting to lower gear.

DOUBLE LEADING BRAKE – A drumbrake assembly with both front shoes self energized during forward wheel rotation.

DOUBLE REDUCTION AXLE – In the double reduction or tripple reduction type final drive, the required speed reduction is obtained in two or more steps. This enables higher torque to be available at the road wheels. In heavy duty and off highway vehicles, multiple reduction is used.

DOUBLE TRAILING BRAKE – A drum brake assembly with both shoes self energized during rearward wheel rotation only.

DRAG – To accelerate a vehicle from standing start, over course one fourth mile in length. Also used by some drivers when referring to challenging another driver to an acceleration race.

DRAG LINK – A steel rod connecting pitman arm to one of steering knuckles. On some installations, drag link connects pitman arm to a centre idler arm.

DRAGSTER – Car especially built for drag racing.

DRAG WHEEL – Special steering wheel used on some dragsters. Often consists of cross bar spoke and portion of rim on each end.

DRIP MOULDING – is a U shaped channel, added to the side rails of the roof panel. It catches water on the roof and direct it to the back of the car during raining.

DRIVE LINE or DRIVE TRAIN – Propeller shaft, universal joints etc. connecting transmission output shaft to axle pinion gear shaft.

DRIVE PINION – A gear in the differential connected to the drive shaft.

DRIVE or PROPELLER SHAFT SAFETY STRAP – A metal strap or straps, surrounding drive shaft to prevent shaft from falling to ground in the event of a universal joint or shaft failure.

DRIVE SHAFT – An assembly of one or two universal joints connected to a hollow tube and used to transmit torque and motion. A shaft in the power train that extends from the transmission to the differential and transmits power from one to the other. Also called PROPELLER SHAFT.

DROP CENTRE RIM – Centre section of rim being lower than two outer edges. This allows bead of tyre to be pushed into lower area on one side while the other side is pulled over and off the flange.

DROPPED AXLE – Front axle altered so as to lower the frame of the vehicle. Consists of bending axle downward at outer ends (solid front axle).

DRUM BRAKE – A brake unit using curved brake shoes which press against the inner circumference of a metal drum to produce braking action.

DUALS – Two sets of exhaust pipes and mufflers one for each bank of cylinders.

DUAL BRAKE SYSTEM – Tandem or dual master cylinder to provide a brake system that has two separate hydraulic systems, one operating the front brakes, the other operating the rear brakes.

DUAL SERVO BRAKES – A drum brake assembly with both front and rear shoes, self energized during forward and rearward wheel rotations.

DUMMY AXLE – provided in some vehicles increases the load carrying capacity of the vehicle. Wheels on the dummy axle can rotate freely.

DUNY BUGGY – Off road vehicle set up to run on sand.

ELLIOT TYPE AXLE – Solid bar front axle on which ends span or straddle steering knuckle.

EMERGENCY BRAKE – is the hand brake, operated by a lever, is used when the vehicle is left parked and prevents the vehicle from moving. The hand brake can be applied to stop the vehicle when the service brake fails.

EPICYCLIC GEAR – In the epicyclic gearing, at least one gear not only rotates about its own axis, but also rotates about some other axis.

EQUALIZER LINK – A common connector in the parking brake system that causes both rear brakes to be applied with the same cable tension.

EVASIVE MANEUVER – Rapid steering changes to avoid obstacles in the path of the vehicle.

EXPANSION TANK – A tank at the top of an automobile radiator which provides room for heated coolant to expand and give off any air that may be trapped in the coolant. Also used in some fuel tanks to prevent fuel from spilling from the tank because of expansion.

FADE (brake) – A condition that occurs when there is little braking effect with full brake pedal force.

FIFTH WHEEL – is the swivelling type arrangement at the rear of a tractor unit. The fifth wheel carries the front part of the semitrailer.

FINAL DRIVE – The final gear reduction between the engine and the drive wheels.

FIXED CALIPER DISC BRAKES – Disc brakes using a caliper which is fixed in position and cannot move.

FLOATING CALIPER DISC BRAKES – Disc brakes using a caliper mounted through rubber bushings which permit the caliper to float, or move, when the brakes are applied.

FLUID FLY WHEEL – A liquid coupling used to transmit the engine effort (torque) to a clutch and transmission. This coupling is always a major part of the engine flywheel.

FOOT PRINT – Area of road that is in contact with the tyre.

FORWARD CONTROL TRUCK – has the engine either in or below the driver's cabin.

FORWARD EFFICIENCY – is the ratio of the amount of driver input torque which is available at the wheels to turn the same to the total amount of input torque from the driver at the steering wheel.

FOUR WHEEL DRIVE – Some cross country vehicles (Jeeps) have this arrangement. In this case, the engine power is transmitted to all the four wheels of the vehicle. The main advantage of this arrangement is the entire vehicle weight is available for traction.

FOUR WHEEL STEERING – Type of steering system in which all the four wheels of a vehicle are turned for steering.

FOUR SPEED TRANSMISSION – A transmission with four forward speeds or gear ratios.

FRAME – The assembly of metal structural parts and channel sections that forms the base and supports the engine and body and is supported by the vehicle wheels.

FRONT AXLE – In a vehicle, the front axle transmits the weight of the front part of the vehicle to the road surface through the front wheels. It also carries the mechanism for steering the vehicle. In the case of front wheel drive, it incorporates both steering and driving mechanisms.

FRONT END GEOMETRY – The angular relationship between the front wheels, wheel attaching parts, and vehicle frame. Includes camber, caster, king pin inclination, toe in and toe out on turns.

FULL FLOATING AXLE – An axle design usually used on heavy trucks where the vehicle weight is carried by bearings in the wheel hubs, or the drive wheels and the axles are used only to transfer driving torque.

GEAR – A wheel with teeth that engage or mesh with teeth of another wheel.

GEAR BOX – A unit which has a series of gears and shafts to vary the speed of the gearbox output shaft compared to the engine speed. This in turn increases the torque and thereby improves acceleration of the vehicle.

GEAR CLASH – A condition in which the gears grind during shifting.

GEAR RATIO – The ratio of the number of teeth on two gears to mesh with each other.

GENERATOR – is the device which converts mechanical energy from the automobile engine into electrical energy. When the generator is sufficiently turned by the engine, it furnishes electrical energy for all the vehicles circuits and replenishes the battery to keep it fully charged.

GRAVITY BLEEDING – A process used to remove air from the brake system using the natural height of the fluid in the master cylinder above the wheel cylinder.

GRIP COEFFICIENT – is numerically equal to the ratio of the force causing uniform wheel slip to the normal road resistance.

HALF AXLE DRIVE SHAFTS – Two in number, transmit the driving torque from the final drive and differential unit to the driving road wheels.

HANDLING – The ease of maneuvering a vehicle without slipping or skidding.

HARSHNESS – Bumpy ride produced by a stiff suspension.

HEEL – Anchor end of a brake shoe.

HOTCHKISS DRIVE – A rear suspension with open propeller shafts with two or three universal joints. Braking torques are transferred to the frame through links, control arms or leaf springs.

HORN RELAY – A relay connected between the battery and horns, when energized by closing of the horn button, it connects the horn to the battery.

HYDRAULIC BRAKE – A brake system using hydraulic fluid, piston and cylinders to provide extremely high pressure for brake application.

HYDRAULIC CLUTCH – A clutch that uses hydraulic pressure to actuate the clutch. Used in heavy duty equipment and where the engine is away from the drivers compartment so that it would be difficult to use mechanical linkages.

HYDRAULIC CONTROL VALVES – A system of valves that senses driving conditions and automatically shifts the transmission.

HYPOID GEARS – Drive pinion and ring gears whose shape allows them to mesh off centre.

IMPENDING SKID – The tyre traction point at which any increase in side or tractive load will produce tyre skid.

INDEPENDENT SUSPENSION – A type of suspension system in which each wheel is independently supported by a spring. A suspension that allows up and down movement of one wheel without affecting the opposite wheel.

INNER TUBE (tyre) – The inside rubber tube assembled in the tyre casing, it maintains the air at sufficient pressure to inflate the casing and adequately support the vehicle weight.

INTEGRAL BODY – has the longitudinal and cross members of the chassis incorporated in the frame work of the load carrying body. With this arrangement, part of the load previously carried by the chassis, is diffused through the body structure.

ISOFLARE – A brake tube flaring method that upsets the tube and allows a strong, secure attachment when used with the proper tube nut and seal.

JOUNCE – A compression load on the springs as the space between the frame and the axle is reduced.

JOUNCE BUMPER – A rubber bumper used to absorb shock during full suspension system movement.

KING PIN – The steel pin on which the steering knuckle pivots, it attaches the steering knuckle to the knuckle support or axle.

KING PIN INCLINATION – Inward tilt of the king pin from the vertical.

KINGPIN OFFSET – is the distance between the centre of the tyre contact patch and intersection of the kingpin or steering axis with the ground. Kingpin offset is also called scrub radius.

KNOCK BACK – Slight axial movement that pushes the caliper pistons into their bore. This causes clearance between the brake lining and the rotor.

KNUCKLE – The part of the suspension that connects the control arms and supports the wheel spindle.

LATERAL LOAD (tyre) – The force on the side of the tyre tread.

LEADING SHOE – A brake shoe that has the drum rotating from the toe toward heel.

LEADING TRAILING BRAKE – A drum brake assembly having one shoe energized in either forward or rearward wheel rotation.

LEAF SPRING – A spring made up of a series of flat steel plates of graduated length, assembled one on top of another.

LEVEL CONTROL (**automatic**) – A suspension system which compensates for variations in load in the rear of the car, positioning the rear at a predesigned level regardless of load.

LIMITED SLIP DIFFERENTIAL – A differential allowing unequal torques to be delivered to the axle shafts.

LINKAGE POWER STEERING – A type of power steering in which the power steering units (power cylinder and valve) are an integral part of the steering linkage.

LORD LEVELING SYSTEM – A system used to level a vehicle that is heavily loaded.

LORD RANGE (**tyre**) – The amount of weight that can be safely carried by a tyre. It indicates the number of plies at which a tyre is rated. Load range B equals 4 ply rating, C equals 6 ply rating, and D equals 8 ply rating.

LUG – The flange stud on an axle or hub on which the drum and wheel are fastened.

MCPHERSON STRUT SUSPENSION – A suspension system in which both wheels are attached to a rigid rear axle housing.

MANUALLY OPERATED TRANSMISSION – A transmission that is shifted from one speed to another by the operator (driver).

MASTER CYLINDER – The liquid filled cylinder in the hydraulic braking system where hydraulic pressure is developed by depression of the brake pedal or movement of the brake lever.

MECHANICAL BRAKES – Brakes operated by mechanical linkage (cables and levers) between the brake pedal and the brakes at the car wheels.

METALLIC BRAKE LINING – A lining having metallic properties used to provide high temperature braking efficiency.

METERING VALVE – A valve that delays pressure build up to the front brakes of a four wheeled vehicle.

MINOR BRAKE ADJUSTMENT – Adjustment of brakes to compensate for brake lining wear.

MODULATOR – A vacuum canister mounted to the outside of the automatic transmission that senses engine load.

MULTIPLE DISC CLUTCH – A clutch that has more than one friction disc, usually there are several driving discs and several driven discs, alternately placed.

NEUTRAL STEER – A vehicle that will maintain the selected turn with no driver input.

NON DIRECTIONAL SENSE – Steering does not lead in any direction.

NON LOAD CARRYING BODY – In this, the loads on the vehicle are transferred to the suspension system entirely by a separate chassis. The body is isolated from the chassis deflection by rubber mountings.

ONE WAY CLUTCH – A clutch that holds in one direction but allows movement in another direction.

OVER DRIVE – A device in the power train of some vehicles that introduces an extra set of gears into the power train. This causes the propeller shaft to overdrive or drive faster than the engine crankshaft. Engine speed is thus reduced without reduction of vehicle speed.

OVER RUNNING CLUTCH – A type of clutch that will transmit rotary motion in one direction only, when rotary motion attempts to pass through in other direction, then the driving member over runs and does not pass motion to the other member.

OVER STEER – The tendency of a vehicle to turn sharper than the turn selected by the driver.

PANHARD ROD – A control rod that connects the frame on one side of the vehicle to the axle housing on the other side to keep the axle housing centred under the vehicle.

PARKING BRAKES – Mechanically operated brakes that operate independently of the (hydraulic) service brakes on the vehicle. They may be set for parking the vehicle or holding the vehicle against rolling. Also called an emergency brake.

PASCAL'S LAW – A principle of hydraulics which states that pressure at any point in a confined liquid is same in every direction and applies equal force on equal areas.

PEDAL BLEEDING – A method of removing air from the hydraulic system parts by applying the brake to raise the pressure in the system to help move the air through the system when the bleeder valves are opened.

PEDAL PULSATION – A rapid up and down movement of the clutch pedal during operation.

PITMAN ARM – That part of the steering gear which is linked to the steering knuckle arms of the wheels; it swings back and forth for steering.

PITMAN ARM STOPS – On some cars (particularly those using linkage power steering), stops are used to prevent excessive pitman arm movement and thus steering linkage movement.

PITMAN SHAFT – The shaft to which the pitman arm is attached in a steering gear.

PLANETARY GEARBOX – A system of gears used in an automatic transmission, a sun gear, planet gears, a carrier and a ring gear.

PLIES – The layers of cord fabric in a tyre carcass, each layer is a ply.

PLYSTEER – The tendency of a tyre to always turn in one direction as it rolls. This is the result of the way the tyre was constructed.

PNEUMATIC TYRES – Tyres that are filled with air to the required pressure.

POWER BOOSTER – A device used to increase the drivers brake pedal force going to the master cylinder, without an accompanying increase in pedal travel.

POWER BRAKE – Conventional brake system that utilizes engine vacuum to operate vaçuum power piston. Power piston applies pressure to brake pedal, or in some cases, directly to master cylinder piston. This reduces the amount of pedal pressure that the driver must exert to stop the vehicle. Also called POWER ASSISTED BRAKE.

POWER RACK – In the saginan power steering unit, a rack that meshes with a sector on the pitman shaft and transmits to the shaft, power from the power cylinder.

POWER STEERING – A device that uses hydraulic pressure to multiply the drivers effort as he turns the steering wheel so that less steering effort is required.

POWER TRAIN or DRIVE TRAIN – The group of mechanisms that carry the rotary motion developed in the engine to the vehicle wheels, it includes the clutch, transmission, drive shaft differential and axles.

PRESSURE CAP (radiator) – A radiator cap with valves that causes the cooling system to operate under pressure and thus at a somewhat higher and more efficient temperature.

PRESSURE PLATE – That part of the clutch which exerts pressure against the friction disc, it is mounted on and rotates with the flywheel.

PREVENTIVE MAINTENANCE – The systematic inspection, detection and correction of failures in a engine, or in a vehicle, either before they occur, or before they develop into major defects.

PRIMARY SHOE – A brake shoe moved by a wheel cylinder to apply the brake.

PROPELLER SHAFT – A shaft in the power train that extends from the transmission to the differential and transmits power from one to the other.

PROPORTIONING VALVE – A valve used to maintain the correct proportion of fluid pressure between the front disc or drum brakes and rear drum brakes. Usually the rear brake pressure is a fraction of front brake pressure.

PUNCTURE SEALING TYRES AND TUBES – Tyres and tubes coated on the inside with a plastic material. Air pressure in the tyre or tube forces that material through holes made by punctures. It hardens on contact with the air to seal the puncture.

RACK AND PINION STEERING GEAR – A steering gear that uses a pinion on the end of the steering shaft which is meshed with a rack on the major cross member of the steering linkage.

RADIAL BIAS TYRE – A tyre in which the plies are laid on radially, or perpendicular to the rim, with a circumferential belt on top of them. The rubber tread is vulcanized on top of the belt and plies.

RADIAL PLY TYRE – Cords running directly across the tyre carcass from bead to bead.

RADIAL SPRING RATE – The amount of radial load required to deflect a tyre (one cm) unit distance.

RADIATOR – In the cooling system, the device that removes heat from the coolant passing through it, it takes hot coolant from the engine and returns the coolant to the engine at a lower temperature. The hot coolant is cooled in the radiator for recirculation.

RADIATOR PRESSURE CAP – The cap placed on the radiator filler tube which pressurizes the cooling system for more efficient operation.

RADIATOR SHUTTER SYSTEM – A system of engine temperature control used mostly on trucks, that controls the amount of air flowing through the radiator by use of a shutter system.

REACTION CONTROL – A feedback mechanism that gives the driver a feel of the amount of input effort being applied.

REAR AXLE ASSEMBLY – A system of gears and axles that transfers power from the drive line assembly to the driving wheels of the automobile.

REAR AXLE RATIO – The ratio between the drive pinion and the ring gear in the differential assembly.

REAR END TORQUE – Reactionary torque applied to the rear axle housing as torque is applied to the wheels; rear end torque attempts to turn the axle housing in a direction opposite to wheel rotation.

REBOUND – An expansion of a suspension spring after it has been compressed as the result of jounce.

RECAPPING – A form of tyre repair in which a cap of new materials is placed on the old tread and vulcanized into place.

RECIRCULATING BALL AND NUT STEERING GEAR – A type of steering gear in which there is a nut (meshing with a gear sector) assembled on a worm, balls circulate between the nut and worm threads.

RELEASE LEVER – In the clutch, a lever that is moved by throw out bearing movement; the movement causes clutch spring pressure to be relieved so that the clutch is released or uncoupled.

RETREAD – A used tyre on which a new tread section is molded.

REVERSE FLUSHING – A method of cleaning a radiator or engine cooling system by flushing in the direction opposite to the normal coolant flow.

RIDE – The characteristic feel as one rides in a vehicle.

RIGID REAR SUSPENSION – A rear suspension system in which both wheels are attached to rigid rear axle housing.

RIM – The metal wheel on which the tyre is mounted.

RING GEAR – A large gear carried by the differential case, meshed with and driven by the drive pinion.

ROAD RESISTANCE – is the resistance of the road surface, which must be overcome when a vehicle travels along the road. This consists of friction between the tyre and road.

ROLLING RESISTANCE – is the resistance caused due to the deformation of the tyres and road, the friction of the tyres on the road surface and friction in the wheel bearings.

ROLL STEER – The steering effect as a result of body lean during a turn.

RUNOUT OF WHEEL – Lack of alignment of wheel or gear to the axle so that the wheel or gear runout or move out of alignment, as wheel or gear rotates.

SAFETY RIM – A type of wheel rim having a hump on the inner edge of the ledge on which the tyre bead rides. The hump helps hold the tyre on the rim in case of blow out.

SCRUB RADIALS – The distance on the road surface under the front tyre between an extension of the pivot axis and the centre of weight.

SCUFF – The tyre slide on the road surface during operation.

SCUFF TRAVEL – The amount of side travel of the tyre as the wheel moves from maximum jounce to maximum rebound.

SEAT ADJUSTER – A device to permit forward and backward (and sometimes upward and downward) movement of the front seat.

SECONDARY SHOE – A brake shoe that is operated by a primary shoe to apply brake.

SELF ADJUSTING BRAKE DESIGNS – Brakes that automatically compensate for wear of the brake linings.

SELF ALIGNING TORQUE – The natural tendency of the tyre to return to the neutral position after being turned.

SEMIMETALLIC BRAKE LINING – A brake lining combining both metallic and organic materials for improved braking performance.

SENSTRONIC BRAKE – CONTROL (SBC) – is basically a brake by wire system which eliminates the need for mechanical linkage between the brake pedal and brake master cylinder.

SEQUENTIAL GEAR BOX – is an electromechanical device that replaces the conventional gear shift mechanism and is bolted to the tunnel section or the floor of the car. It converts the conventional floor shifter to an electronically activated sequential shift system that is electronically controlled by microprocessors.

SERIES – The designation of a tyre aspect ratio.

SERVICE BRAKE SYSTEM– The main braking system of the vehicle which controls braking effect proportional to the drivers demand.

SHACKLE – Swinging support by which one end of a leaf spring is attached to the vehicle frame.

SHIM – A slotted strip of metal used as a spacer to adjust front end alignment on many cars and to make small correction in the position of the body sheet metal and other parts.

SHIMMY – Rapid oscillations, in wheel shimmy, for example, the front wheel tries to turn in and out alternately and rapidly (a violent front wheel shake). This causes the front end of the car to oscillate or shimmy.

SHOCK ABSORBER – The assembly on the vehicle that checks excessively rapid spring movement and oscillations. A device placed at each vehicle wheel to regulate spring rebound and compression.

SHOE – The part of a brake that supports the lining.

SHORT LONG ARM SUSPENSION – A suspension system in which a long and a short control arms are used to support the wheel.

SINGLE LEADING SHOE – A drum brake having two shoes; one is leading another is trailing. Leading shoe tends to wedge itself into the brake drum and provides more braking action than the trailing shoe.

SINGLE REDUCTION AXLE – In the single reduction type final drive, the required speed reduction (say up to 7 to 1) is obtained in one step.

SIPES – Slits in the tyre tread to produce more blade surface for traction.

SKID – A tyre sliding on the road surface.

SKID CONTROL – A device that operates to prevent wheel lock up during braking and thus skidding.

SLIDING MESH GEAR BOX – The gear box consists of three shafts and a set of gears, gear selector mechanism and gear shift lever. Different gears are engaged by sliding the appropriate gears.

SLIP ANGLE – The angle between the tyre and the actual directional movement.

SLIP JOINT – In the power train, a variable length connection that permits the drive shaft (propeller shaft) to change its effective length.

SOCIETY OF AUTOMOTIVE ENGINEERS (SAE) – A professional engineering society responsible for setting many vehicle standards used in the world today.

SPIN BALANCER – A car tyre balancer which turns the raised tyre at a road speed where imbalance can be detected by wheel shake.

SPINDLE (steering system) – A part of the steering knuckle assembly on which the front wheels are mounted.

SPONGY PEDAL – A brake pedal that feels soft or spongy.

SPRING FREQUENCY – Springs are also compared in frequency. Springs which return quickly to their original shape or oscillate rapidly after being deflected are said to have higher frequency than those that return or oscillate slowly.

SPRING RATE – The flexibility of a spring depends on the property called spring rate. It is defined as the weight in kg required to deflect it one cm. A soft spring has a lower rate than a stiff or inflexible spring.

SPRING SHACKLE – Provides a means for the leaf spring assembly to compensate for changes in its length.

SPRING SUSPENSION – The operating components of a spring suspension system, which absorbs the force of road shocks by flexing and twisting.

SPRUNG WEIGHT – That part of the vehicle which is supported on springs (the frame and body for example).

SQUIRM – The twist of the tyre tread in the foot print.

STABILIZER SHAFT – An interconnecting torsion bar between left and right lower suspension arms on a vehicle which reduces body roll on turns and adds stability.

STAR WHEEL – An adjustable link between the primary and secondary brake shoes.

STEER ANGLE – The angle the wheels are turned to from straight ahead.

STEERING AND IGNITION LOCK – A locking device that locks the ignition switch in the off position, and also locks the steering wheel so that it cannot be turned.

STEERING ARM – The arm attached to the steering knuckle to turn the knuckle, and wheel, for steering.

STEERING AXIS – The centre line of the ball joints in a front suspension system extended to the road surface.

STEERING AXIS INCLINATION – The inward tilt of the steering axis or front wheel pivot from the vertical.

STEERING COLUMN – The housing that supports the steering shaft.

STEERING GEAR – That part of the steering system, located at the lower end of the steering shaft, that carries the rotary motion of the steering wheel to the vehicle wheels for steering.

STEERING KICKBACK – Sharp rapid movements of steering wheel as the front wheels encounter obstructions in road, the shocks of these encounters kickback to the steering wheel.

STEERING KNUCKLE – The front wheel spindle or stub axle which is supported by the king pin, so that it and the wheel can be turned for steering. Part of the front suspension system.

STEERING LINKAGE – Linkage that connects the steering gears to the front wheels.

STEERING RATIO – The number of degrees the steering wheel is turned divided by the number of degrees the vehicle wheels are turned.

STEERING SHAFT – Shaft extending from steering gear to the steering wheel.

STEERING STOPS – limit the angular deflections of the front wheels. They avoid rubbing of tyres against the frame or against the fenders which would cause undue wear and tear of the tyres.

STEERING SYSTEM – The mechanism that enables the driver to turn the wheel axles (usually the front) and thus turn the wheels away from the straight ahead position so that the vehicle can be guided.

STEERING WHEEL – The wheel at the top of the steering shaft in the drivers compartment which is used to guide or steer, the vehicle.

STONE SHIELD – is bolted to the radiator support and the fenders. It fits beneath the bumper of the car. It prevents striking of small flying stones on to the radiator grille and radiator and thus avoids their damage.

STOP LIGHT SWITCH – A switch applied by the master cylinder pressure that turns on brake lights.

STOPPING DISTANCE – is the distance in which a vehicle will be brought to rest from steady speed, when the brake is applied.

STREAMLINING – The shaping of an object that moves through a medium (such as air or water) or past which the medium moves, so that less energy is lost by parting and reuniting of the medium as the object moves through it.

STUMBLE – The term related to vehicle driveability, the tendency of an engine to falter and then catch, resulting in a noticeable stumble effect felt by the driver.

SUSPENSION – The suspension system supports the vehicle body and at the same time isolates the vehicle and its occupants from shocks and vibrations generated by the road surface. It also maintains steering control and stability at all times.

SUSPENSION ARM – In the front suspension, one of the arms is pivoted at one end to the frame and at the other to the wheel (steering knuckle) support.

SUSPENSION COMPLIANCE – Rearward and upward movement of the suspension when the tyre meets an obstacle on the road surface.

SUSPENSION GEOMETRY – The angular action of the suspension as it goes from its static position to the extremes of travel (compared to vertical lines).

SYNCHROMESH – A device in the transmission that synchronizes gears about to be meshed so that there will not be any gear clash. Also called SYCHRONIZER.

TACTILE SENSOR – A sensor that allows the vehicle operator to feel when a certain condition is reached. Disc brake pads are made to vibrate when worn to the point where replacement is necessary and this vibration is felt in the brake pedal.

TANDEM MASTER CYLINDER – is the unit installed in some large cars and commercial vehicles have a split hydraulic system with two separate cylinders and reservoirs in the master cylinder. This avoids the possibility of complete brake failure due to a fracture in the pipe line leading to one brake cylinder.

THROWOUT BEARING – Bearing operated by the clutch linkage used to disengage the clutch.

TIE RODS – In the steering system, the rods that link the pitman arm to the steering knuckle arms.

TILT STEERING WHEEL – A type of steering wheel which can be tilted at various angles, due to a flex joint in the steering shaft.

TOE – The leading edge of the brake shoe. The angle between the centre lines of the front wheels.

TOE IN – The turning in of the front wheels, wheels are closer together at the front than at the back of the wheels.

TOE OUT – The turning out of the front wheels, where wheels are farther apart at the front than at the back of the wheels.

TOE OUT DURING TURNS – Difference in angles between the two front wheels and the car during turns. Inner wheel, in a turn, turns out or toes out more. Also called steering geometry and cornering wheel relationship.

TORQUE CONVERTOR – A device in the power train consisting of three or more rotating members. It transmits power from the engine through a fluid to the reminder of the power train and provides varying drive ratios with speed reduction and increase in torque.

TORQUE DRIVE TRANSMISSION – A transmission similar to the power glide but lacking the self shifting ability.

TORQUE TUBE DRIVE – The type of rear suspension in which the torque tube surrounding the propeller shaft absorbs the rear end torque.

TORSIONAL LOAD – Loads on the brakes and suspension caused by torque.

TORSION BAR SPRING – A long, straight bar, fastened to the frame at one end and to a suspension part at the other.

TRACKING – The following of the rearwheels, directly behind, or in the tracks of, the front wheels.

TRACTIVE FORCE – The friction force in the contact patch that causes torque on the wheel.

TRAILING SHOE – A brake shoe with its anchor at the toe end.

TRAMP – Up and down motion or hopping of the front wheels experienced at higher speeds due to unbalanced wheels or to excessive wheel run out. Also called high speed shimmy.

TRANSAXLE – A drive assembly combining the transmission and final drive assemblies in one casing.

TRANSFER CASE – A unit located at the back of the regular gear box, in the four wheel drive arrangement. A pinion fixed to the gear box shaft, drives a wheel in the transfer case. The driven wheel in the transfer case has a differential which distributes the drive equally between the front and rear axles.

TRANSMISSION – The device in the power train that provides different gear ratios between the engine and rear wheels, as well as reverse.

TRANSMISSION DRAIN PLUG – A plug at the bottom of the transmission to drain the lubricant.

TRANSMISSION FILLER PLUG – A plug on the side of the transmission used to add transmission lubricant.

TRIM HEIGHT – Specified level, vehicle height above the road surface.

TRIPLE POINT JOINT – A universal joint using bearings on three axes to maintain a constant plane of drive, making it a constant velocity joint.

TUBED TYRE – Inside the tyre, there is an endless tube fitted with a valve. Air is forced through the valve and is retained inside the tube under pressure. The air acts as the cushioning medium.

TUBELESS TYRE – A tyre that has the air sealed between the rim and tyre and does not use an inner tube.

TURNING RADIUS – The relative angles of the two front wheels during a turn.

TWO DISC CLUTCH – A clutch having two friction discs for additional holding power used in heavy duty equipment.

TYRE – The casing and tube assembled on a vehicle wheel to provide pneumatically cushioned contact and traction with the road.

TYRE BEAD – The inner reinforced edge of a tyre that hold it to the wheel rim.

TYRE CARCASS – The main structural part of the tyre to which tread rubber is attached.

TYRE CONTACT PATCH – The part of a tyre that contacts the road surface making a footprint.

TYRE FOOT PRINT – The area on the road in contact with the tyre.

TYRE FORCE VARIATION – Changes in the tyres radial spring rate as it rolls under radial loads.

TYRE SLIP – A slight tyre slide while making a turn.

TYRE ROTATION – Changing the position of tyres on the automobile to evenout the amount of wear.

TYRE RUNOUT – The amount the tyre wobbles as it rotates.

TYRE SERIES – The groupings of tyre sizes having the same aspect ratio.

TYRE TREAD – is that part of the tyre that is designed to run on the road surface. The tread rubber is grooved with a pattern that will provide maximum friction force, (which provides good traction and reduces the possibility of skidding) and minimum noise.

UNDER STEER – The tendency of the vehicle not to turn as much as the wheels are turned.

UNITIZED CONSTRUCTION – A type of automobile body and frame construction in which the frame and body parts are welded together to form a single unit.

UNSPRUNG WEIGHT – That part of the vehicle which is not supported on springs (the wheels and tyres for example). The vehicle weight moved by variations in the road surface.

UNIVERSAL JOINT – The part of the drive line assembly that allows for a change in angle of the drive line as the vehicle goes over bumps.

VACUUM BRAKE – is the device in which the braking effect is due to the difference of pressures that acts on the opposite sides of a diaphragm. In this unit, one side of the piston or diaphragm is exposed to atmospheric pressure while the other side to a pressure which is below the atmospheric pressure.

VARIABLE RATE SPRINGS – provide a low rate for ordinary service and higher rate for heavy obstruction or loads. These consist of a conventional spring and below which is placed a small auxiliary spring with several leaves. Under heavy loads, the auxiliary or helper spring strengthens the main spring more and more as the main spring is compressed.

VARIABLE RATIO STEERING – A steering gear that provides a different ratio during parts of a turn.

WADDLE – A sideways vehicle shake due to a faulty radial tyre. Most noticeable when a vehicle moves slowly.

WANDER – A condition in which the vehicle does not follow a straight path and randomly drifts in one direction or the other.

WEIGHT TRANSFER – The changes in radial loads on the front and rear wheel tyres due to the centre of gravity location ring braking.

WHEELS – The wheels (wheel and tyre assembly) support the weight of the vehicle. The assembly provides ride quality, load carrying capacity, and vehicle handling characteristics.

WHEEL ALIGNMENT – The position of the front wheels in relation to the suspension and steering geometry.

WHEEL BALANCER – A device that checks a wheel, either statically or dynamically, for balance.

WHEEL BASE – Distance between center of the front wheel and center of rear wheels.

WHEEL CYLINDER – In the hydraulic braking system, hydraulic cylinders placed in the brake mechanisms at the wheels; hydraulic pressure from the master cylinder causes the wheel cylinders pistons to move the brake shoes.

WHEEL FIGHT – The tendency of a steering system to be easily deflected by uneven road surfaces. Causes changes in toe that result in tyre wear.

WHEEL OFFSET – The distance between wheel attachment flange and the wheel rim centre plane.

WHEEL PACK BEARING – A preassembled self-lubricated bearing assembly used on the drive wheels with independent suspension.

WHEEL SIDEWAYS DISPLACEMENT – Sideways movement of the wheel as the suspension goes from jounce to rebound.

WHEEL SIZES – are indicated by three measurements, namely rim diameter, rim width and flange height.

WHEELSLIP – Sideways movement of the tyre tread across the foot print.

WHEEL RUNOUT – The amount the wheel wobbles as it rotates.

WHEEL TRAMP – Tendency of the wheel to move up and down so it repeatedly bears hard or tramps, on the pavement. Sometimes called high speed shimmy.

WIND SHIELD WIPER – A mechanism which utilizes a rubber blade to wipe the wind shield, it is either vacuum or electrically operated.

CHAPTER 25

AIR COMPRESSORS

ABSOLUTE HUMIDITY – Actual quantity of water vapour in the air, usually expressed as so many grains of moisture in a cubic foot of air.

ABSOLUTE PRESSURE – Pressure measured from the true zero or point of no pressure.

ADIABATIC COMPRESSION – Compression of air without receiving or giving up heat.

AFTER COOLER – A type of surface heat exchanger in which compressed air is cooled after compression.

AIR – A gas consisting of mechanical mixture of 23.2% (by weight) of oxygen 75.5% nitrogen and 1.3% argon, 21 % (by volume) of oxygen, 78.06% nitrogen and 0.94% argon.

AIR COMPRESSOR – A machine (driven by any prime mover), which compresses air into a receiver to be used at a greater or shorter distance.

AIR COOLED COMPRESSOR – A compressor whose cylinder has cast integral numerous thin fins to form excess cooling surface exposed to a draught of cool air which forms the medium to carry off some of the heat of compression.

AIR ENGINE – A very small reciprocating engine driven by compressed air.

AIR EXHAUSTER – A suction fan, a vacuum pump.

AIR METER – An apparatus used to measure the rate of flow of air or gas.

AIR RECEIVER – A vessel into which compressed air is discharged, to be stored until required.

ANEMOMETER – An instrument for measuring the velocity of flow of a gas, either by mechanical or electrical methods.

ATMOSPHERIC PRESSURE – The force exerted by the weight of the atmosphere on every point with which it is in contact.

AXIAL COMPRESSOR – A multistage, high efficiency compressor comprising alternate rows of moving and fixed blades attached to a rotor and its casing respectively. The flow of fluid is essentially parallel to the axis of the compressor.

BOYLE'S LAW – At constant temperature, the absolute pressure of a gas varies inversely as its volume.

CENTRIFUGAL COMPRESSOR – A compressor designed to deliver large quantities of air or gas at low pressure, moved by centrifugal force generated by a fast revolving rotor.

CHARLE'S LAW – At constant pressure, the volume of a gas is proportional to its absolute temperature. At constant volume, the pressure is proportional to its absolute temperature.

CLEARANCE VOLUME EFFECT – Volumetric efficiency of the reciprocating compressor depends upon the clearance volume in the air cylinder. The greater the clearance volume, the greater the volume of the cylinder occupied by the clearance air which expands and prevents the entrance of free air during the early part of the admission stroke.

COMPRESSED AIR – Air forced into a smaller space than it originally occupied. When air is compressed both its pressure and temperature rise.

COMPRESSION CONSTANT – According to Boyle's law, product of pressure and volume at any instant is constant, at constant temperature.

COMPRESSION EFFICIENCY – Ratio of the theoretical power required to compress the amount of air actually delivered to the actual power developed in the cylinder as shown by the indicator diagram.

COMPRESSOR OVERALL EFFICIENCY – Ratio of actual power developed in the air cylinder as shown by the indicator diagram to the power supplied to the compressor shaft.

CONSERVATION OF ENERGY – Energy can be transmitted from one body to another or transformed in its manifestations, but can neither be created nor destroyed.

DIRECT CONNECTED COMPRESSOR – A compressor in which the prime mover is attached direct to the compressor without any interposed transmission such as chain, belt etc.

DISPLACEMENT OF COMPRESSOR – The volume displaced by the net area of the compressor piston. This is the capacity of an air compressor, usually expressed in cum per minute.

DOUBLE ACTING COMPRESSOR – A reciprocating compressor in which compression occurs on either side of the piston during every stroke.

ECCENTRIC AND STRAP – An eccentric is a disc having its axis of rotation out of its centre. It is equivalent to a crankpin which is so large in diameter that embraces the shaft to which it is attached and dispenses with arms. Converts rotary motion into reciprocating motion.

EFFICIENCY – Ratio of the useful work performed by a prime mover to the energy expended, that is, the output divided by the input.

ENBLOCK CYLINDERS – Two or more cylinders cast integral, that is all in one casting.

FEATHER VALVE – A valve which consists of a strip of ribbon steel which covers a slightly narrower slot when the value is closed.

FINGER VALVES – Valves consist of narrow strips of stainless steel, fastened to the seat at one end and free to flex along their length. Suitable for light service.

FIXED COMPRESSOR – A compressor mounted upon a permanent base as concrete for service not requiring removal from place to place.

FREE AIR – Air at atmospheric condition at the point where a compressor is installed.

FREE AIR UNLOADER – An automatic device that varies the amount of air or gas being pumped.

HOPPER COOLED SYSTEM – A non (external) circulating system. The cylinder has an open water jacket of considerable volume.

INDICATED HORSE POWER – The actual power developed within a cylinder as calculated from the indicator diagram.

INDICATED WORK – The work of compression plus the work of expulsion of the air from the cylinder minus the workdone on the piston by the pressure of the air during admission.

INLET LINE LOADER – An unloader that automatically opens and closes the inlet line under pressure variations in the receivers.

INTERCOOLER – A type of surface heat exchanger placed between two cylinders of a two stage compressor so that heat of compression generated in the first stage cylinder may be removed (in part or whole) from the air as it passes through the intercooler to the second stage cylinder.

ISOTHERMAL COMPRESSION – Compression of air at constant temperature. Law of compression is PV= constant.

KINETIC ENERGY – Energy due to momentum, that is, the energy of a moving body, which is equivalent to saying, dynamic inertia.

MECHANICAL EQUIVALENT OF HEAT – Relationship between the unit of heat and unit of work.

MEAN EFFECTIVE PRESSURE – The average resultant pressure acting on the piston during the stroke, that is the effective pressure which compresses and discharges the air. This is the difference between the mean forward pressure and the mean back pressure.

MECHANICAL EFFICIENCY – The ratio of the air indicated horse power to the brake horse power supplied to the compressor shaft.

MULTISTAGE COMPRESSION – Dividing the compression of air into two or more stages so as to obtain the work saving due to a nearer approach to isothermal compression by using intercooling.

PISTON SPEED – The total distance travelled by the piston in one minute, not the actual velocity at any given instant.

PORTABLE COMPRESSOR – A small compressor which is easily moved from place to place.

POWER – The rate at which work is done, that is work divided by the time in which it is done. Unit of power is horse power equal to 4500 mkg/ minute.

POWER DRIVEN COMPRESSOR – A compressor having a separate prime mover and connected by a suitable transmission such as a belt.

PRIME MOVER – An apparatus or mechanism whereby motion and force are received directly from some natural source of energy (fuel) and transmitted into some motion by means of which the power may be conveniently applied.

R (gas constant) – An experimentally determined constant which is equal to the mechanical work done by the expansion of unit weight of a perfect gas at a constant pressure while heat is added to increase its temperature by one degree centigrade.

RATIO OF COMPRESSION – Ratio of final volume to the initial volume during compression.

RECIPROCATING COMPRESSOR – A compressor having a piston which is made to move to and fro, that is forward and backward and thereby compresses and delivers air.

RELATIVE HUMIDITY – Degree of saturation of the air with water vapour as determined by the use of the wet and dry bulb thermometers.

RESISTANCE – The quality of not yielding to force or external pressure, that quality of a body which acts in opposition to the pressure of another.

ROTARY COMPRESSOR – A compressor having a vane rotor or its equivalent mounted eccentrically in a stationary casing.

SEMI FIXED COMPRESSOR – A unit larger than the portable type where skids are used in place of being mounted on a truck, the adaptation being for service where frequent moving is not necessary.

SEPARATOR – Device through which the compressed air after being cooled in the after cooler, is sent so as to separate the moisture from the air by centrifugal force.

SINGLE ACTING COMPRESSOR – A reciprocating compressor in which compression takes place on one side of the piston during alternate strokes.

SINGLE STAGE COMPRESSOR – A compressor in which the compression cycle takes place in a single cylinder.

SLIPPAGE EFFICIENCY – The ratio of volume of air actually measured to the apparent volume accounted for by the indicator diagram.

SPECIFIC HEAT OF AIR – Amount of heat that is to be supplied to raise the temperature of 1 kg of air through 1 degree C.

SPECIFIC HEAT OF AIR AT CONSTANT PRESSURE – Total specific heat of air which is made up of (1) the internal work of raising the temperature of air, and (2) the external work of pushing away the atmosphere to make room for its expansion.

THROW OF THE ECCENTRICITY – Twice the eccentricity or the amount of reciprocating motion produced.

TWO STAGE COMPRESSOR – A compressor in which compression begins in one cylinder and is completed in the second cylinder. It divides the compression range between the two cylinders and permits cooling between the cylinders.

TWO STAGE CYLINDER – A cylinder of special construction with a step piston, the low pressure being at the top while the high pressure is formed around the trunk.

VALVE GEAR – The mechanism or combination of parts by which a reciprocating or to and fro motion is imparted to the valve from the rotary motion of the shaft.

VOLUMETRIC EFFICIENCY – The ratio of the actual number of cubic meter of free air (at 1.03 kscm abs and 15° C) compressed per unit of time to the number of cubic meter of piston displacement during that time.

WATER COOLED COMPRESSOR – A compressor whose cylinder is water jacketed and through which flows a current of cold water which functions as a transmission medium to carry off some of the heat of compression.

WORK – The overcoming of resistance through a certain distance by the expenditure of energy.

CHAPTER 26

REFRIGERATION AND AIR CONDITIONING

ABSOLUTE HUMIDITY – The weight of the water vapour which is associated with unit quantity of air.

ABSORBER – A device for absorbing a refrigerant, a low side element in an absorption system.

ABSORPTION REFRIGERATION SYSTEM – One in which the refrigerant, as it is absorbed in another liquid, maintains the pressure difference needed for successful operation of the system.

ABSORPTION REFRIGERATOR – A plant in which ammonia is continuously evaporated from an aqueous solution under pressure, condensed, allowed to evaporate (so absorbing heat), and then reabsorbed.

ABSORPTIVITY – Ability of a material to absorb heat.

ACCUMULATOR – A steel shell partly filled with liquid refrigerant, the space above which is maintained by the compressor at a pressure corresponding to the required refrigerant temperature. The shell is placed in a suction line for separating liquid entrained in the suction gas.

ACTIVATED AMMONIA – Desiccant which operates by adsorption of water molecules. A form of aluminium oxide AlO_2.

ACROLEIN – A warning agent having the formula CH_2 CH CHO is often used with methyl chloride to call attention to the escape of refrigerant. The material has a compelling, pungent odour and causes irritation of the throat and eyes.

ACTUATORS – Secondary control mechanisms which function in response to the requirements of the primary group in actually controlling some part of the refrigeration system.

ADIABATIC COMPRESSION – Compression of a vapour or gas in such circumstances that there is insufficient time for any substantial exchange of heat between it and its surroundings.

ADIABATIC COOLING – Method in which paramagnetic salts are precooled, and then demagnetized, thereby producing further cooling.

AERATION – A term generally employed with reference to air circulation or ventilation. In milk cooling, it refers to a method where the milk flow over refrigerated surfaces is exposed to the atmosphere.

AIR CIRCULATION – A method for natural or forced motion of air.

AIR CONDITIONING – The simultaneous control of all, or at least the first three of the following factors affecting the physical and chemical conditions of the atmosphere within a structure: Temperature, humidity, motion, distribution, dust, bacteria, or lesser degree human health or comfort.

AIR CONDITIONING UNIT – Equipment designed as a specific air treating combination, consisting of means for ventilation, air circulation, air cleaning and heat transfer with control means for maintaining temperature and humidity within prescribed limits.

AIR COOLED CONDENSERS – Condenser used to cool the refrigerant, the cooling effect depends on air drawn through tubes and fins for a good distribution of air.

AIR COOLER – The cold accumulator used in the Linde process of air liquefaction for the preliminary cooling of the air.

AIR DUCTS – Pipes or channels through which air is distributed throughout building or machinery for heating and ventilation.

AIR INFILTRATION – The inleakage of air through cracks and crevices and through doors, windows or other openings, caused by wind pressure or temperature difference.

AIR LIQUEFIER – A type of gas refrigerating machine based on the STIRLING CYCLE, the cycle of the hot air engine.

AIR WASHER – An enclosure in which air is forced through a spray of water in order to cleanse, humidify or dehumidify the air.

ALFOL – Technical name for thin corrugated aluminium foil in narrow strips, used for heat insulation, for which it is effective by reason of the numerous small air cells formed when packed.

AMBIENT SENSOR – A temperature sensor that provides an outside air temperature signal for an automatic temperature control type air conditioning system.

AMBIENT TEMPERATURE – In a domestic or commercial system having an air cooled condenser, it is the temperature of the air entering this condenser.

AMMONIA-NH_3 – One of the earliest compounds used as a refrigerant.

AMMONIA MACHINE – An abbreviation for a compression refrigerating machine using ammonia as a refrigerant. Similarly, freon, sulphur dioxide machine etc.

ANALYSER – Device used in the high side of an absorption system for increasing the concentration of vapour entering rectifier or condenser.

ANTIFREEZE LIQUID – A substance added to the refrigerant to prevent formation of ice crystals at the expansion valve.

ATMOSPHERIC CONDENSER – A condenser operated with water which is exposed to the atmosphere.

AUTOMATIC EXPANSION VALVE – A pressure actuated device which regulates the flow of refrigerant from the liquid line into the evaporator to maintain a constant evporator pressure.

AUTOMATIC REFRIGERATION SYSTEM – One which regulates itself to maintain a definite set of conditions by means of automatic controls and valves usually responsive to temperature or pressure.

BASIC REFRIGERATION CONTROL – Device that starts, stops, regulates and/ or protects the refrigeration system and its components.

BAUDELOT EVAPORATOR – An open type of cooler in which the liquid to be cooled flows from distributing troughs or headers over a cooling surface consisting of sets of grids or a pair of stamped corrugated metal sheets forming channels.

BLEEDER – A pipe sometimes attached to a condenser to lead off liquid refrigerant, parallel to the mainflow.

BRINE – Any liquid cooled by the refrigerating system and used for the transmission of heat.

BRINE SYSTEM COOLING – Any system whereby brine, cooled by a refrigerating system, is circulated through pipes to the point where the refrigeration is needed.

BUTANE – A hydrocarbon, flammable refrigerant used to a limited extent in small units.

CALCIUM CHLORIDE – A chemical having the formula $CaCl_2$ which is in granular form, is used as drier. Soluble in water.

CALCIUM SULPHATE – A solid chemical of the formula $CaSO_4$ which may be used as a drying agent.

CAPACITY – In a refrigerating machine, it is the heat absorbing capacity per unit time, usually measured in ton or kcal/hr.

CAPILLARY – A tube with a very small inside diameter, its diamcter and length control the flow of the refrigerant; dividing point between the high side and the low side of the system.

CARBON – One of the elements used in refrigeration.

CARBONDIOXIDE – One of the earliest compounds used as a refrigerant.

CARBON TETRA CHLORIDE – A liquid having the formula CCl_4 (also known as carbona) which is non -inflammable solvent used for removing grease and oil and loosening sludges.

CENTRIFUGAL COMPRESSOR – A compressor in which the necessary increase in the pressure of the refrigerant vapour being obtained by imparting a high velocity to it by the rotation of an impeller.

CHANGE OF STATE – A change from one state to another as from liquid to solid, from liquid to gas etc.

CHARGE – The amount of refrigerant in a system.

CHARGING CYLINDER – A cylindrical container for refrigerant that has a calibrated sight glass so that the mechanic can measure the flow of refrigerant into the air conditioning system.

CHLORINATED FLUORO CARBON – The chemical family into which air conditioning refrigerants such as refrigerant 12 fall.

CIRCUITS – The flow of a refrigerant through separate rows of tubes rather than through one single tube.

CLEARANCE VOLUME EFFICIENCY – Ratio of the weight of the refrigerant circulated by a compressor having no losses except that due to clearance, to the weight circulated by a perfect machine.

CLOSED SYSTEM – Chilled water from the flash tank is pumped through a coil to cool air and is then returned to the flash tank.

COEFFICIENT OF PERFORMANCE – The ratio of the refrigerating effect to the heat equivalent of the indicated horse power of the refrigerant compressor.

COIL – Any cooling element made of pipe or tubing.

COLD STORAGE – A trade or process of preserving perishables on a large scale by refrigeration.

COMFORT CHART – A psychrometric chart; strictly a chart showing the effective temperatures.

COMPOUND COMPRESSION – In compound or multistage compression, the refrigerant is compressed through part of the pressure range in one compressor (or in one stage of a multistage compressor) and then passed to a second compressor, or stage, of smaller swept volume, which carries the compression further.

COMPOUND GAUGE – A typical low pressure test gauge, which has a scale that indicates both pressure and vacuum.

COMPRESSOR – A device that takes a refrigerant vapour at a low temperature and pressure and compresses it to a lower volume and thereby raises it to higher temperature and pressure.

COMPRESSOR CRANKSHAFT SEALS – Prevent air from entering the compressor, and oil and refrigerant from escaping.

COMPRESSION RATIO – The ratio of two pressures, the absolute discharge pressure divided by the absolute suction pressure.

COMPRESSION SYSTEM – A refrigerating system in which the pressure imposing element is mechanically operated.

CONCENTRATORS – Evaporate excess water from brine which has been diluted by melted ice and frost.

CONDENSATION – Process by which a vapour is changed into a liquid without changing temperature. Condenses the hot, high pressure refrigerant vapour from the compressor to a warm, high pressure liquid which flows to the receiver dehydrator.

CONDENSER (general) – That part of the refrigeration system in which the refrigerant condenses and in so doing gives off heat.

CONDENSER DUTY – Amount of heat transferred in a given time from the refrigerant to the cooling medium in the condenser.

CONSTANT PRESSURE VALVE – An automatic expansion valve that holds the pressure at a constant level regardless of the load.

CONSTANT TEMPERATURE VALVE – A valve responsive to temperature of thermostatic bulb, of the throttling type, located in suction line of an evaporator to reduce refrigerating effect on coil to just maintain a desired temperature.

CONTAINER CAPACITY – The ability of a container to hold the material the quantity of material which may safely be contained in a container.

COOLING UNIT – A specific air treating combination consisting of means for air circulation and cooling within the prescribed temperature limits.

COOLING WATER – Water used for condensation of the refrigerant.

COPPER PLATING – Formation of a film of copper usually on compressor walls, pistons or discharge valves.

CRYOGENICS – Science of producing and applying temperature below –250° F.

CRYOGENIC SUPER CONDUCTOR SYSTEM – Uses helium to cool conductors to within a few degrees of absolute zero where they offer no electrical resistance.

CRYOHYDRATE – A eutectic brine mixture of water and any salt, mixed in proportions to give the lowest temperature.

CYCLE OF REFRIGERATION – A complex course of operation of a refrigerant back to the starting point, measured in thermodynamic terms, also used in general for any repeated process for any system.

CYCLING CLUTCH CONTROL SYSTEM – One in which the compressor is run intermittently to maintain a desired temperature.

DEGREE DAY – A unit, based upon temperature difference and time, used in specifying the nominal heating load in winter.

DEFROSTING – Removal of accumulated ice from the cooling unit.

DEFROSTING CYCLE – A cycle which permits cooling unit to defrost during off period.

DEFROSTING EVAPORATOR – Unit in which frost accumulates on cooling coils when the compressor operates and melts after the compressor shuts off.

DEHUMIDIFY – To remove water vapour from the atmosphere, to remove water or liquid from stored goods.

DEHYDRATOR – A device used to remove moisture from the refrigerant.

DESICANT – Material used in a drier to trap moisture from the refrigerant. Also called drying agent.

DEWAR FLASK – A container which consists of alternate layers of radiation shields and spacer material in high vacuum.

DEW POINT – That temperature at which the air (space) becomes saturated with water vapour. When the air is cooled to the dew point, water vapour can condense into liquid form (provided its latent heat is removed).

DEW POINT THERMOSTAT – A thermostat used in such a way as to control humidity.

DICHLORODIFLUORO METHANE – The chemical compound known as Freon 12 or R -12.

DIELECTRIC – Thin insulating material separating two conductor plates in a capacitor.

DIFFERENTIAL – Difference of temperature or pressure between the on and off operation of the control.

DIRECT EXPANSION – A system in which the evaporator is located in the material or space refrigerated or in air circulating passages communicating with such space.

DIRECT EXPANSION EVAPORATOR – One that contains only enough liquid to continue boiling as heat is absorbed by it.

DISCHARGE SHUT OFF VALVE – A manual valve installed on the compressor, which controls the flow of the refrigerant from the cylinder head of a compressor to the discharge line.

DISPLACEMENT, ACTUAL – The volume of gas at compressor inlet conditions actually moved in a given time.

DISPLACEMENT, THEORETICAL – The total volume displaced by all the pistons of a compressor for every stroke during a definite interval, usually measured in cubicmetre per minute.

DOMESTIC REFRIGERATOR – A refrigerator for home use.

DOUBLE ACTION PISTON COMPRESSOR – A type of compressor used in automotive A/C systems in which an axial swash plate pressed to the shaft is used to drive the pistons.

DOUBLE EFFECT EVAPORATOR – An arrangement of two evaporators such that the vapour from No. 1 is carried over into the tubes of No. 2. The drain from the first evaporator is led to the hot well and the drain from the second evaporator is led to the fresh water condenser and there cooled down to be used by the passengers.

DOUBLE THICKNESS FLARE – Indicates that the flare thickness of a tube end is made up of two thicknesses of tubing.

DRIER – Device designed to remove moisture from a refrigerant.

DRIERITE – Desicant which operates by chemical action.

DRY BULB TEMPERATURE – The actual temperature of the air as measured by an ordinary thermometer.

DRY ICE – Frozen carbondioxide, sold under the trade names, such as CARDICE and DRICOLD. This has the property of passing directly from the solid to the gaseous state without becoming a liquid.

DRY TYPE EVAPORATOR – An evaporator of the continuous tube type where refrigerant from a pressure reducing device is fed into one end and the suction line is connected to the outlet end.

EBULLATOR – A device inserted in flooded evaporator tubes to prevent evaporator from being oil bound.

EJECTOR – A device which utilizes static pressure to build up a high fluid velocity in a restricted area to obtain a lower static pressure at that point so that fluid from another source may be drawn in.

ELECTROMAGNETIC CLUTCH MECHANISM – Device which when engaged, turns the compressor shaft to start piston movement.

ELECTROPNEUMATIC AIR CONDITIONING SYSTEM – A system in which an adjustable sensing device permits the selection of comfortable conditions, has two thermistors to monitor and sense both the outside air and the air in the passenger compartment.

ELIMINATOR PLATES – Protect refrigerated spaces and air from brine spray.

EMULSIFICATION – Formation of an emulsion, *i.e.*, a mixture of small droplets of two or more liquids which do not dissolve with each other.

EQUALIZERS – Connections used with thermostatic expansion valves when the superheat setting of the expansion valve cannot control the amount of refrigerant which flows through the coil.

EQUILIBRIUM – Condition existing at saturation, the molecules of the refrigerant in liquid state are changing into the vapour state as rapidly as vapour molecules are changing into the liquid state.

EUTECTIC POINT – The lowest freezing point (temperature obtainable) when the concentration of solid in a solution is increased gradually.

EUTECTIC SOLUTION – A solution which can be made so that it freezes and melts at a specific temperature.

EVACUATE – To discharge refrigerant from the air conditioning system into the atmosphere or a holding tank, and then to vacuum pump the system in order to boil away any moisture.

EVAPORATOR – Device in the low pressure side of a refrigeration system through which the unwanted heat flows; absorbs the heat in the system in order that it may be moved or transferred to the condenser.

EVAPORATOR (automotive ac system) – Device that cools, dehumidifies, and takes the pollen and dust from the air before it enters the passenger compartment.

EVAPORATOR DUTY – The amount of heat which can be removed by the evaporator *i.e.*, the amount of refrigeration accomplished.

EVAPORATOR PRESSURE (temperature) – CONTROL VALVE SYSTEM – Uses either a suction throttling valve, a pilot operated absolute valve, or an evaporator pressure regulator valve to control evaporator temperature.

EVAPORATOR REGULATOR VALVES – Provide independent temperature control for each evaporator.

EXPANSION VALVE – Metering device which provides a restriction so that there is a steady flow of refrigerant and also maintains the difference of pressure required to change the state of the refrigerant.

FLARING – Method of forming or preparing the ends of tubing to connect them directly with or through the use of fittings.

FLEXING DISC VALVE – One type of valve commonly used in compressors. It is a one way valve.

FLOODED EVAPORATOR – One that is full of liquid refrigerant at all times. Additional liquid is permitted to enter only to replace that which boils away.

FREEZER BURN – Surface damage due to excessive drying during freezing.

FREON 12 – Refrigerant used in automatic air conditioners. Also known as Refrigerant -12 and R-12.

FROST HEAVE – Refers to the movement of ground as a result of being frozen because of insufficient insulation underneath a cold store.

FUSES – Devices used for protection of electrical circuits, either cartridge or plug type.

GAS VOLUME CONTROL – Used to regulate the amount of gas needed to produce certain desired temperatures and conditions in domestic absorption automatic control refrigerators.

GROUNDING – Protection against static charges which sometimes build upon operating equipment.

HALIDE LEAK DETECTOR – Operates on acetylene to detect vapour leaks of halogen refrigerants.

HEAT OF RESPIRATION – Heat given off by cargo.

HELIUM LIQUEFIER REFRIGERATOR – The complete system for liquefying helium.

HERMETIC COMPRESSOR – The compressor unit in which the motor and the compressor are manufactured as a single self contained unit housed within a casing, the electric motor is in contact, therefore with the refrigerant.

HIGH PRESSURE CUTOUT – A pressure operated switch which stops the machine on the rise of pressure to a level approaching danger, and usually has to be reset by hand.

HIGH SIDE FLOAT – Metering system which locates the float and needle valve on the high pressure side of the refrigeration system.

HOLD OVER PLATES – Containers that hold the eutectic and provide refrigeration.

HUMIDITY – Moisture in air.

HUMIDISTAT – Operating control which reacts to variation in humidity.

ICE MAKING CAPACITY – Ability of a refrigerating system to make ice, starting with water at room temperature.

INDICATED HORSE POWER – Rate at which work is usefully expended in the compressor *i.e.*, actually utilized in compressing the refrigerant vapour and expelling it from the compressor.

INDUCTIVE DEVICE – Designed to convert electrical energy to magnetic and then to mechanical energy.

INSULATION – Any material that effectively slows down the transfer of heat.

INSULATORS – Materials that normally deter the flow of electrons.

LATENT HEAT – Heat energy which causes a change of state without any change of temperature.

LIMITING CONTROLS – Safety controls.

LIQUID LINE CHARGING VALVE – Used for high side charging.

LIQUID LINE SHUT OFF VALVE – Manual valve installed in liquid line near condenser well to shut off flow of refrigerant between the condenser and the liquid line.

LIQUID NITROGEN SHIELD – At room temperature, it can absorb any heat leakage from outside and reduce temperature between itself and cryogenic refrigerant surrounding the cable.

LIQUID NITROGEN SYSTEM – A non-mechanical refrigeration system for transport use.

LITHIUM BROMIDE – Used in combination with water in absorption cooling systems.

LOW PRESSURE CONTROL – An electric switch responsive to pressure, connected into the low pressure part of a refrigeration system. Usually closes at high and opens at low pressure.

LOW PRESSURE CUTOFF SWITCH – Senses system pressure only, wired in series with the magnetic clutch.

LOW SIDE FLOAT – Metering system which locates a float in the low pressure side of the refrigeration system.

LOW TEMPERATURE TRANSPORT – Refrigerated trucks that maintain temperatures in the range of 0°C and below.

MANIFOLDING – In direct expansion or dry evaporators, the method of circulating the refrigerant through separate rows of tubes.

MANUAL CONTROL SYSTEM – An A/C system in which the driver of an automotive vehicle selects heating and cooling by use of a lever which mixes warm and cold air to desired temperature.

METERING DEVICES – Restrict the flow of the refrigerant from the high to the low side, regulate the flow of the refrigerant according to the needs of the system.

METHYLENE CHLORIDE (CH_2Cl_2) – A halogenated hydrocarbon which is considered a safe refrigerant.

MODULATING CONTROLS – Provide for variations by steps as contrasted to the off and on operation of the refrigeration systems with ordinary controls.

MODULATING THERMOSTAT – Used to operate dampers on DX coils and valves for varying the flow of chilled water.

MODULATING THERMOSTATIC EXPANSION VALVE – Varies the capacity of the valve in response to variations in load on the system.

MOISTURE VAPOUR SEAL – A tight barrier placed outside the insulation to prevent pushing of moisture through the insulation by vapour pressure.

MULTIPLE UNIT INSTALLATION – One in which two or more evaporators in different refrigerators are operated from one compressor, or vice versa.

NONFLEXING RING PLATE TYPE VALVE – One type of valve commonly used in compressors.

NONFROSTING EVAPORATORS – Use only the thermostatic expansion valve type of refrigerant control, operate at a temperature close to freezing.

NONMECHANICAL REFRIGERATION SYSTEMS – Those that obtain the required high and low pressures by some method other than a mechanical compressor.

OIL FAILURE RELAY – An oil pressure switch inserted in the compressor lubricating system and wired to shut down the machine in the event of an oil failure.

OIL SEPARATOR – Device used to separate oil from the refrigerant gas, returning the oil to the compressor and allowing the refrigerant to continue on its circuit through the refrigerating system.

OPEN SYSTEM – Chilled water is sprayed into the air to be cooled and it is then collected in the air washer tank and returned to the flash tank and is again cooled.

OPERATING CONTROLS – Sensitive to changes in the desired conditions such as temperature (or its related pressure) and humidity.

pH VALUE – Logarithm to base 10 of the reciprocal of the concentration of hydrogen ions. Measure of the acidity or alkalinity of a solution.

PRESSOSTAT – An automatic switch connected by a small bore tube to the suction of the compressor, stopping the latter when the pressure falls to a certain value and starting it again after a definite rise in pressure.

PRESSURE RELIEF VALVE – Used to minimize the possibility of explosion when air temperature surrounding a refrigeration system may rise to a point where it causes the pressure of the refrigerant gas to increase to a danger point.

PRIMARY REFRIGERANT – A substance used as the working fluid in the vapour compression cycle, as distinct from a secondary refrigerant, which in some cases is used as an intermediate conveyor of heat between the substance to be cooled and the primary refrigerant.

PSYCHROMETRIC CHART – A graph, the coordinates of which are usually either dry bulb temperature and absolute humidity, or enthalpy (total heat) and absolute humidity. Families of lines are there showing constant dry bulb temperatures, wet bulb temperatures, enthalpies, absolute humidities, relative or percentage humidities and specific volumes.

PURGING – A method of removing air and moisture from a refrigerating system by means of the refrigerant gas pushing some of the air ahead of it and out of the system.

REFRIGERANT – Substance which is circulated in a refrigeration system to transfer heat.

REFRIGERANT FAMILY – Safest group of refrigerants produced by manipulating the atoms of carbon tetrachloride with those of fluorine and hydrogen.

REFRIGERATING CAPACITY – The ability of a system to remove heat as compared with the cooling effect produced by the melting of ice, expressed as a rate of heat removal, kcal/ hr or tons/ 24 hrs.

REFRIGERATION SYSTEM – The part of the refrigeration/ air conditioning system that includes compressor, condenser, evaporator, control valves and switches and tubing. It absorbs the heat from air in the duct housing/space and transfers it to the outdoors.

RELATIVE HUMIDITY – The percentage of moisture in the air compared with the maximum amount that the air can hold, at the prevailing temperature. Also expressed as the ratio of the pressure of the water vapour present to the maximum possible water pressure at the prevailing temperature.

RESISTIVE CRYOGENIC SYSTEM – Uses nitrogen to cool the electrical conductors to temperatures where their electrical resistance is very low.

REVERSE CYCLE REFRIGERATION – Uses rejected heat to produce warmth.

ROTARY COMPRESSOR – Compressor which consists of a cylindrical casing containing a shaft whose axis is eccentric to that of the cylinder, the shaft carries a rotor having radial slots in which blades slide and the tip of the blades press against the casing by their inertia or with the aid of springs.

SAFETY HEAD – Unit which is not fixed to the top of the cylinder but is held down by heavy springs and allows the passage of liquid refrigerant or oil, which it does by lifting bodily whenever the pressure in the cylinder becomes abnormally high.

SEALED UNIT – The compressor unit in which the compressor, usually rotary, and the driving motor are contained within a welded steel shell.

SECONDARY REFRIGERANT – Chilled liquid like water which is circulated to distant units where the air is to be cooled in individual rooms.

SEMI HERMETIC COMPRESSOR – The compressor unit in which the motor and compressor are a single unit, but the motor is detachable from the compressor and therefore capable of field repairs.

SHELL AND COIL CONDENSER – A cylindrical shell, usually vertical, containing a water coil, and within the shell the refrigerant is condensed.

SHELL AND TUBE EVAPORATOR – An evaporator having a cylindrical casing, containing a number of tubes through which the liquid to be cooled flows and the refrigerant is contained within the casing.

SHRADER VALVE – A spring loaded valve through which a connection can be made to a refrigeration system, also used in vehicle tyres.

SIGHT GLASS – In a car air conditioner, a viewing glass or window set in the refrigerant line, usually in the top of the receiver dehydrator, the sight glass allows a visual check of the refrigerant passing from the receiver to the evaporator.

SILICA GEL – Desicant which operates by adsorption of water molecules.

SINGLE THICKNESS FLARE – The part of the tubing that forms the flare is the thickness of the tubing.

SPRAY HEADER – Perforated pipe mounted along the ceiling of the cargo compartment of a transport.

STEAM JET SYSTEM – Uses a device in which the extremely rapid flow of a vapour through a narrow tube reduces the pressure and permits evaporation of a liquid, produces a cooling effect.

STRONG AQUA – An ammonia and water solution with a concentration of almost 30 per cent ammonia, used in ammonia absorption cooling system.

SUCTION LINE – Runs from evaporator to compressor, returns the heat laden gases from the evaporator.

SUCTION SERVICE VALVE – Manual shut off valve installed on the compressor. Also called suction valve.

SUCTION THROTTLE VALVE – The compressor is in continuous operation and the valve is opened and closed by sensing the actual evaporator operating pressure.

SULPHURDIOXIDE – An old refrigerant that was used to recharge units.

SUPERHEATING – The rise in temperature resulting from the addition of heat to the refrigerant vapour either in the evaporator or in the suction line.

SUPERHEAT SWITCH – Designed to protect the A/C system compressor against damage when the refrigerant charge is partially or totally lost.

SUPER INSULATION – Alternate layers of radiation shields and spacer material operating in a high vacuum.

SURGE CHAMBER – A drum or container into which liquid enters from the metering device in order to recirculate the refrigerant in a flooded evaporator.

SWAGING – A means of shaping copper tubing so that two pieces may be joined without the use of a fitting.

TEMPERATURE CONTROL – An electric switch responsive to temperature of thermostatic bulb or element.

TERMINAL DEWARS – Insulated containers used to prevent heat transfer and permit the individual conductor phases to be connected into thermally and electrically graded pot head assemblies.

THERMAL LIMITER FUSES – Designed to protect the A/C system compressor against damage when the refrigerant charge is partially or totally lost.

THERMOELECTRIC REFRIGERATION – Depends upon passing electrical energy to a couple through two dissimilar semiconductors.

THERMOPNEUMATIC AIR CONDITIONING SYSTEM – System which uses the mechanical principles of the thermostat to monitor vacuum motors which adjust the air valves and switches.

THERMOSTAT – An automatic switch, the opening and closing of which is actuated by change of temperature.

THERMOSTATIC EXPANSION VALVE – Control valve which maintains constant superheat in the evaporator, also used for the temperature control, operates on increased pressure resulting from a rise in temperature. Also called THERMOSTATIC VALVE.

THROTTLING VALVE – Dampens fluctuations of pressure gauge and provides a way to close off the port entirely.

TON REFRIGERATION – A ton refrigeration (TR) is that rate of removal of heat which would transform water at 0°C into ice at the same temperature at the rate of one ton in every 24 hours.

VACUUM PUMP – A device used to evacuate systems in preparation for charging them with a refrigerant.

VALVE RETAINER – A device which limits the lift of the valve.

WATER CONTROL VALVE – Used in A/C systems to regulate the flow of coolant to the heater core.

WATER ICE REFRIGERATION SYSTEM – Heat is absorbed as ice melts, producing the desired cooling effect.

WET BULB TEMPERATURE – The temperature read, by a wet bulb thermometer, this is an ordinary thermometer the bulb of which is wetted by being surrounded by a sheath of muslin kept wet by pure water.